Full STEAM Ahead

Recent Titles in
Libraries Unlimited Professional Guides for Young Adult Librarians
C. Allen Nichols and Mary Anne Nichols, Series Editors

Teen-Centered Library Service: Putting Youth Participation into Practice
Diane P. Tuccillo

Booktalking with Teens
Kristine Mahood

Make Room for Teens! Reflections on Developing Teen Spaces in Libraries
Michael G. Farrelly

Teens, Libraries, and Social Networking: What Librarians Need to Know
Denise E. Agosto and June Abbas, Editors

Starting from Scratch: Building a Teen Library Program
Sarah Ludwig

Serving Teen Parents: From Literacy Skills to Life Skills
Ellin Klor and Sarah Lapin

Teens Go Green! Tips, Techniques, Tools, and Themes in YA Programming
Valerie Colston

Serving Latino Teens
Salvador Avila

Better Serving Teens through School Library–Public Library Collaborations
Cherie P. Pandora and Stacey Hayman

Teen Games Rule! A Librarian's Guide to Platforms and Programs
Julie Scordato and Ellen Forsyth, Editors

Dragons in the Stacks: A Teen Librarian's Guide to Tabletop Role-Playing
Steven A. Torres-Roman and Cason E. Snow

Cooking Up Library Programs Teens and 'Tweens Will Love: Recipes for Success
Megan Emery Schadlich

FULL STEAM AHEAD

Science, Technology, Engineering, Art, and Mathematics in Library Programs and Collections

Cherie P. Pandora and Kathy Fredrick

Libraries Unlimited Professional Guides
for Young Adult Librarians Series
C. Allen Nichols and Mary Anne Nichols, Series Editors

LIBRARIES UNLIMITED™

An Imprint of ABC-CLIO, LLC
Santa Barbara, California • Denver, Colorado

Library of Congress Cataloging-in-Publication Data

Names: Pandora, Cherie P., author. | Fredrick, Kathy, author.
Title: Full STEAM ahead : science, technology, engineering, art, and mathematics in library programs and collections / Cherie P. Pandora and Kathy Fredrick.
Description: Santa Barbara, California : Libraries Unlimited, [2017] | Series: Libraries Unlimited professional guides for young adult librarians series | Includes bibliographical references and index.
Identifiers: LCCN 2017043550 (print) | LCCN 2017032888 (ebook) | ISBN 9781440853418 (ebook) | ISBN 9781440853401 (paperback : acid-free paper)
Subjects: LCSH: High school libraries—Activity programs—United States. | Middle school libraries—Activity programs—United States. | Young adults' libraries—Activity programs—United States | School librarian participation in curriculum planning. | School libraries—Collection development—United States. | Science—Study and teaching—United States. | Technology—Study and teaching—United States. | Engineering—Study and teaching—United States. | Arts—Study and teaching—United States. | Mathematics—Study and teaching—United States.
Classification: LCC Z675.S3 (print) | LCC Z675.S3 P2156 2017 (ebook) | DDC 025.5/678—dc23
LC record available at https://lccn.loc.gov/2017043550

ISBN: 978-1-4408-5340-1 (print)
 978-1-4408-5341-8 (ebook)

21 20 19 18 17   1 2 3 4 5

This book is also available as an eBook.

Libraries Unlimited
An Imprint of ABC-CLIO, LLC

ABC-CLIO, LLC
130 Cremona Drive, P.O. Box 1911
Santa Barbara, California 93116-1911
www.abc-clio.com

This book is printed on acid-free paper ∞

Manufactured in the United States of America

To the female role models in my life: to Aunt Dodo and Aunt Marilynn who showed me how to be a strong, independent, working woman and to my Mom who taught me everything.

—*Cherie*

To Carl, whose dedication to writing set a high bar for my work, much love and endless thanks for your forbearance.
To John and Chris, my dear sons, and daughter-in-law Brittany, who have always made me strive to do the best I can: love always and forever.
To my siblings and their families, who knew me from the start. We've come a long way from that family farm in rural northern Wisconsin. Thanks for your love and support.

—*Kathy*

CONTENTS

SERIES FOREWORD

For those of us who advocate for teens and work with them in libraries, we know how vital it is to prepare them with skills and tools to thrive in the future economy. If you are a public or school librarian, odds are you have been following the STEAM initiative as it gains strength. Teens benefit from engaging in STEAM activities that involve problem-solving, creativity, experiential learning, and collaboration. The skills they can develop through this application and innovation will serve them well in the future. As a librarian with limited time and budget, it is easy to become overwhelmed with all of the possibilities of incorporating STEAM initiatives in your library's plan of service. Cherie Pandora and Kathy Fredrick have years of experience successfully working with teens. They have done the work for you! This guide will put you on a path to STEAM success. While many STEAM programs start with the idea for a makerspace, the authors show there can be so much more. You will find activities and resources, but other topics include funding and support as well as meeting curriculum standards, evaluation and measurement, and even professional growth in relation to STEAM.

We are proud of our continued association with Libraries Unlimited/ABC-CLIO, as they are committed to publishing practical quality works for library employees working with teens.

We hope you find this book, as well as our entire series, to be informative, providing you with valuable ideas as you serve teens and that this work will further inspire you to do great things to make teens welcome in your library. If you have an idea for a title that could be added to our series, or would like to submit a book proposal, please e-mail us at bittner@abc-clio.com. We'd love to hear from you.

Mary Anne Nichols
C. Allen Nichols

ACKNOWLEDGMENTS

My thanks to the kind and generous librarians who have taught me, mentored me, supported me, and been my safety net through times good and bad. My thanks to Krista Taracuk; to Book Club members Janet Rowland, Joanna McNally, Marie Sabol, Diane Dillon, Lori Guerrini, Jane Puleo, and Sue Subel; and to the Library La-Las Jennifer Schwelik, Terri Fredericka, Mary Rawlings, Paula Deal, Kathi Redrup, Diane Deibel, and Kaethe Buchholz. And with special thanks to Kathy Fredrick, a superb writer who was willing to take on this adventure with me.

Cherie

I have been fortunate to learn from wonderful colleagues around the world; librarians and educators who have provided me with support and inspiration that enriched my career and my practice. Thank you. Special thanks to Jo Weyrauch, Julie and Dick Tennie, Diane Dillon, Marti Franks, and Mary Strauss. And many thanks to Cherie Pandora, who may not have realized what a challenge she presented to me when she wondered if I'd be interested in a project she was considering. Thanks for your support, expertise, and unfailing generosity throughout.

Kathy

INTRODUCTION: THE POWER OF STEM AND STEAM

[Science] is more than a school subject, or the periodic table, or the properties of waves. It is an approach to the world, a critical way to understand and explore and engage with the world, and then have the capacity to change that world . . .
—President Barack Obama, "Remarks at White House Science Fair," 2015

Extend this quotation to STEM, and we have the elements to grow scientists, engineers, inventors, and citizens who can make real-world connections with science and technical subjects. STEM aims to refocus education to develop these qualities for all students: a way to approach the world, a focus on critical thinking, a willingness to explore and engage, and involvement with real-world experiences. The growth of STEM is predicated on these elements. STEM is recognized through the four disciplines represented by the letters of the acronym: *S*cience, *T*echnology, *E*ngineering, and *M*athematics. (Note: the *A* for *A*rts will be addressed later in this introduction.) The actual definition of STEM goes far beyond the limits of each subject discipline, as educators look for active learning leading to higher student engagement. Traditionally, science and mathematics were taught totally independent of the other, so that content knowledge existed in silos. The STEM approach is to bring the subjects together so that students build understanding and apply concepts across disciplines (Gerlach, 2012). STEM encourages work across disciplines, with a project-based approach that creates authentic learning experiences. It is the interaction between these fields that can broadly extend young people's understanding as they move from school to community as productive employees and citizens.

How was it that the concept called STEM came into being? It grew out of the realization by business, government, and education leaders that we, as a nation, are not preparing enough students for a future ready career in STEM fields. Data bears this out. Program International Student Assessment (PISA) results ranked the United States 35th in mathematics and 27th in science out of 64 countries tested (Kent, "5 Facts about American Students," 2015). This discrepancy led to much discussion of what was needed to reverse this situation. In their STEM agenda for K-12 education, the National Governors Association noted a need for additional educators in STEM fields and a need for more consistent standards from state to state as key issues moving forward (National Governors Association, 2011, 6).

In the workplace, demand for workers skilled in STEM fields has out-paced the number of candidates available. The Committee on STEM Education of the National Science and Technology Council reported to the president that between 2000 and 2010, the growth in STEM jobs was three times greater than in non-STEM jobs (Committee, 2011, 1). Looking to the future, STEM-related jobs are projected to grow to more than nine million between 2012 and 2022, an increase of one million jobs over 2012 employ-ment levels (Vilorio, 2014, 3). In technology fields, the Bureau of Labor Statistics projects that between 2014 and 2024 there will be a demand for 135,300 software developers and 118,600 computer systems analysts. The growth rate for both jobs is much faster than average with 17 percent growth rate for software developers and 21 percent growth rate in jobs for computer system analysts (Bureau of Labor Statistics, 2015).

Beyond the economic realities spurring interest in STEM, educators are also looking at how best to incorporate STEM in schools. There is a rec-ognition that each element of STEM should be integrated into the school program in an interdisciplinary approach where there are connections between the subject areas. This led to the development of STEM as a cur-riculum, as a methodology, and as an educational practice.

What does STEM look like in schools today? The American Society for Engineering Education published a study in 2014 that identified com-ponents of a STEM program. It is focused on student engagement, with authentic activities relevant to student experience. There is greater focus on applying the concepts learned in science and mathematics. Engineering is a core application to provide hands-on activities, applying a design process and implementing practical components to conceptual knowledge from science and mathematics. This focus on applying knowledge adds soft skills in communication and teamwork for successful learning experiences. To accomplish this, pedagogy shifts to inquiry, with a student-centered focus (Jolly, "Stem vs. STEAM," 2014).

What can be seen in a STEM classroom? Look for interactive lessons, where students are working in groups to solve problems. The most dramatic change is in the addition of engineering to school curricula. There are far more engineering classes in schools, where science and math concepts are integrated through use of a design process to solve real-world problems. Fabrication labs (aka fab labs) offer stations where students move from project to project in teams. In science classrooms, technology tools provide a way for students to collect and analyze data and run simulations, combining scientific concepts with technology literacy.

Then along came STEAM. As more schools adopted the STEM approach and made changes in programming, arts educators began advocating for the addition of *A*, the arts, to STEM and the STEAM acronym was born. Since STEM looks for hands-on, practical experiences and creative thinking, arts of all kinds can enrich the work students do. Steven Ross Pomeroy pointed out that Nobel laureates in the sciences are often also involved in the arts, and that noted psychiatrist Carl Jung developed an archetype of the artist-scientist for those who engineer, build, and invent (Pomeroy, blog, August 22, 2012). Making connections between scientific and artistic enterprises gives students another way to approach STEM. Adding artistic endeavors and projects to programs and lessons allows students to view STEM through a much more open-ended, creative lens; it also allows us to tap another talented group of young adults, those that enjoy the visual and performing arts of music, art, theater, and dance.

Where does this leave the library? Conceptually, the shift to inquiry, to project-based learning experiences, and interdisciplinary activities match how libraries approach research and instruction with students. Professional school librarians can incorporate STEM through collaborative work with teachers, in library-based activities, and as a part of clubs and promotions. As a key player in professional development efforts, professional librarians can work on initiatives that help colleagues and administrators to make STEM and STEAM work. The library can also be an informal learning center, so that we can offer young adults a variety of hands-on projects, allowing them to create products as a fun, free-time activity. This can provide rich experiences for those already interested in STEM, as well as lures for those who would not identify as self-proclaimed science or math geeks. We can enhance the perceptions of STEM and STEAM for all students through this work.

If you have not been involved in STEM work in your school or library, it is time to find out what is happening. Many schools participate in outside-school events like coding days, robotics competitions, and science fair opportunities. Learn more about what your school is involved in, and offer support. Your facility can be a location to use after-school, helping to organize activities, and getting resources to support these efforts. The librarian does not have to set up these activities alone. Collaborating with other libraries and librarians in public and academic settings, with community agencies, museums, and with local businesses gives breadth and depth to your knowledge, your programs, and your resources. Find mentors and partners who can contribute to programming and build connections for young people.

On a more formal level, instructional planning is a key part of the school librarian's role. If science-related fields are not a particular strength for you as a librarian, build familiarity by reviewing and understanding content standards in a variety of disciplines. Review resources and find out about STEAM programs that are already in operation. Thanks to the engineering strand, STEAM uses a design process model that parallels the research models used by librarians in instructional activities. Curation of information and resources is a librarian's forte, and can be a way to begin your work with STEAM. Use the library's online presence and other social media tools to bring resources together for educators and families in your school and district to use. Be a resource for teachers; be a resource for the students as they work on STEM projects.

Public librarians can support students in exploring STEM through activities and materials in their own collections and in partnering with school librarians to reach out to teens and pre-teens. Compelling programming that focuses on connecting students with hands-on activities can lure students into fields and interests they might not have considered. The growth of the maker movement has led to makerspaces in many libraries. This is a natural fit for STEAM, most importantly in providing hands-on materials and experiences for all students. In programming, bringing in STEAM professionals to talk about their careers can start a connection that could lead to opportunities like career shadowing, workplace visits, and mentoring. Regular meeting times for groups who enjoy coding, strategic gaming, and robotics can help students connect with others who share their interests.

This book is designed to be a STEM/STEAM guide for librarians who work with middle and high school students in schools and public libraries. The primary focus is identifying resources—online, print, people, organizations—that can be used for collection development, collaborative instructional planning, programming, identifying funding sources, and discussing professional development opportunities. Each STEAM area—science, technology, engineering, art, and mathematics—will be covered in chapters that include a discussion of the discipline and how the library can become an integral partner in STEAM education. We also include resources, ideas, and agencies that can help us to encourage girls and minorities to be more involved with STEM programs.

Our world is increasingly complex and technology creates connections that were not possible a generation ago. As librarians, we focus on helping young people grow their skills, whether reading or making or getting that project just right. We nurture learners, and grow their skillsets. We have a key role to play in establishing and supporting STEM and STEAM efforts in our schools and communities, a key role in helping students become future ready citizens of the world.

FURTHER READING

American Youth Policy Forum. "Understanding STEM Education: A Discussion of the Key Issues, Efforts and the Role of Federal Policy; A Special Briefing for Congressional Staff." January 27, 2012. http://www.aypf.org/wp-content/uploads/2012/05/STEM%20Facts%20and%20Resources%20Handout.pdf

Killeen, Erlene Bishop. "Supporting STEM to Remain Relevant." *Teacher Librarian* 43, no. 2 (December 2015): 52.

Knowledge Quest. STEM/STEAM Blog Archive. http://knowledgequest.aasl.org/category/blogs/stem/

Pittinsky, Todd L. and Nicole Diamante. "Going Beyond Fun in STEM." *Phi Delta Kappan* 97, no. 2 (October 2015): 47–51.

Vazquez, Jo Anne. "STEM Beyond the Acronym." *Educational Leadership* 72, no. 4 (December 2014/January 2015): 10–15.

REFERENCES

Bureau of Labor Statistics. "Most New Jobs," December 17, 2015. http://www.bls.gov/ooh/most-new-jobs.htm

Committee on STEM Education. National Science and Technology Council. *Federal Science, Technology, Engineering and Mathematics (STEM) Education 5-Year Strategic Plan*. May 2013. https://www.whitehouse.gov/sites/default/files/microsites/ostp/stem_stratplan_2013.pdf

Gerlach, Jonathan. "STEM: Defying a Simple Definition." *NSTA Web Digest, NSTA Reports.* April 11, 2012. http://www.nsta.org/publications/news/story .aspx?id=59305

Kent, Lauren. "5 Facts about American Students." Pew Research Center Fact Tank. Pew Research Center, August 10, 2015. http://www.pewresearch.org/ fact-tank/2015/08/10/5-facts-about-americas-students/

National Governors Association. "Building a Science, Technology, Engineering and Math Education Agenda: An Update of State Actions." December 2011. http://www.nga.org/files/live/sites/NGA/files/pdf/1112STEMGUIDE .PDF

Obama, Barack. "Remarks by the President at White House Science Fair," March 23, 2015, In U.S. Department of Education. "Science, Technology, Engineering and Math: Education of Global Leadership." http://www.ed.gov/stem

Pomeroy, Steven Ross. "From STEM to STEAM: Science and Art Go Hand-in-Hand." *Scientific American.* Blog. August 22, 2012. http://blogs.scientific american.com/guest-blog/from-stem-to-steam-science-and-the-arts-go-hand-in-hand/

U.S. Department of Education. "Science, Technology, Engineering and Math: Education of Global Leadership, Five-Year Strategic Plan for STEM Education." http://www.ed.gov/stem

Vilorio, Dennis. "Stem 101: Intro to Tomorrow's Jobs." *Occupational Outlook Handbook* (2014): 2–12. http://www.bls.gov/careeroutlook/2014/spring/art01 .pdf

1

◇ ◇ ◇

GROWING A STEAM PROGRAM

The beauty of STEAM programming is that it can help young people make connections between disciplines and encourage creativity across fields. Creating programs that touch on multiple STEAM areas can be a great benefit to your library and to your young adult (YA) patrons. If you have previously created programs in one content area, this is your chance to pull in additional resources—teachers, patrons, students, and community—to expand your program into new areas. This chapter offers suggestions for establishing a program, enhancing your current offerings, and growing your program. This effort can begin with stand-alone sessions focused on specific STEAM topics that align most closely with efforts that may already be underway and can grow as your capacity for new activities, planning, and implementation allow.

The joy of providing STEM opportunities through the library program lies in becoming a hub for an integrated approach to STEM, blending science, math, and technology elements in activities and programs. In schools, this allows students to explore without as large a funding commitment of staff, lab space, and scheduling that could overload students in full-blown coursework. Efforts in the library can be a proving ground for curricular changes and can be a parallel to classroom activities that engage students who may not select a particular course.

Library programming in STEM/STEAM can also provide for students who want to dabble rather than go into a full course. Informal learning setting may also be less daunting to those underrepresented in STEM classes—girls and minorities. Key to program development may be collaboration with a variety of community resources: libraries, museums, and businesses with STEM prerequisites. They can be resources in developing programs.

PLANNING AND IMPLEMENTATION

There are plenty of steps involved in program planning and implementation. Begin with a review of where things stand at present using an environmental scan to look at external and internal trends and needs. Once the scan is complete, determine goals for programming. It is good to develop your own ideas about what can happen in and through the library, so that you can articulate your goals and vision to others. At some point, a group of interested parties (STEAM discipline teachers, interested students, parents) can be convened to review and affirm or adapt these goals. As goals are being developed, keeping administrators or supervisors in the loop is critical. No one likes to be surprised, and your efforts may be better appreciated with this transparency. Next develop goals for the program, develop the plan of action, and determine ways to monitor the plan and build sustainability.

Environmental Scan

Understanding your situation in terms of existing issues, programming, and activities in the school, library, and community is good preparation for planning a program. An environmental scan can be a systematic way of reviewing current status and identifying both opportunities and problems. This involves looking at the assets already in place to benefit such programming, delineating the challenges to be faced, and mapping out what actions and allies will be needed to proceed.

In determining needs, begin by outlining what you already know. Determine what STEM initiatives may already be under way in the school or district. Check with administrators or supervisors, as well as colleagues who would be likely candidates for STEAM activities. Identify the players who can be enlisted as allies and information sources. In the public library, technology staff may have programs that touch on STEAM areas (gaming, coding, robotics). In schools, there may be committees already working on STEM implementation. Investigate whether you could become part of such a group, or at least sit in on discussions to get a sense of the direction being taken for STEM and STEAM.

Within the school or library, consider getting numbers to back up plans. Is there interest in after-school or lunchtime sessions around STEM activities? Develop questionnaires or interest surveys to collect information from colleagues and from students. These should be separate versions. Keep in mind, particularly for students, that they will need a description of what such activities might involve as you develop questions. If the school or library has protocols for surveys, make sure to follow those guidelines and make appropriate arrangements to gather information.

Think about the library facility itself and what might be needed for STEAM activities. Is there workspace for activities? Where can materials be stored? If a makerspace is one goal, how will this impact other areas of the library? What changes might need to be made in the physical layout of the library? What resources might be needed?

Goal Setting

Plan the work and work the plan. Once information is gathered in the environmental scan, the next step is to clarify goals for the programming that is being contemplated. What are the desired outcomes that will provide maximum benefit to students? How will you know that goals have been met? What will success look like? This can be a good collaborative process with any partners involved (other teachers, community members, and other library staff members). State goals clearly, and make them realistic.

One widely used practice for goal setting is known by the acronym SMART. Figure 1.1 identifies what each letter can represent (there are a variety of versions). The second column describes what each element brings to the goal setting process.

Specificity will keep goals from being too amorphous. If the goal can't be pinned down to something specific, more information may be needed to build understanding of the underlying issues. For instance, increasing student engagement is admirable. But what does this mean specifically: engagement in class participation, presentations, and clubs? How could this be rewritten? For example, this might be the initial goal: young people will have fun building a robot. A SMART version of this goal would be: 15 students will create a robot from a kit in a two-hour session and be able to describe the process they used to complete the project to each other and to adults facilitating the session. There are a specific number of students and a specific project time bound with an attainable outcome in reporting out to the group.

S	Specific, significant, stretching	Broad statements not allowed – specify what is involved in the goal.
M	Measurable, meaningful, motivational	What specific ways will be used to determine if the program or any part of it are effective?
A	Attainable Achievable Action-oriented	Build for success, with steps that can be accomplished in addition to goals that may stretch efforts over time.
R	Results based Realistic Relevant	It is not just setting something up that makes an accomplishment; it would be the percentage of students impacted, the rate of growth, etc. Use specifics, like percentages, numbers,
T	Time-bound Trackable Tangible	Have a timeline, and be as realistic as possible. Timelines can always be adjusted as goals are monitored, and their use provides an impetus for action.

Figure 1.1: SMART Goals Chart

Clarity on goals will make it easier to articulate the program going forward. The goals can be used to form an action plan, outlining the steps to take, with a timeline for each step. Answer the "what, where, when, and who" questions for each idea to turn your plan into actionable steps.

- What is the activity or program step, and the resources needed?
- Where will it be held/accomplished?
- When will it be completed?
- Who will be involved?

The "why" should be answered in the goal statements, which form the rationale for any action plan. For more information on developing an action plan, see Chapter 11, Evaluation and Measurement, of this book.

One important part of developing a new program is monitoring what is being accomplished, so goals can be adjusted along the away. Be realistic in how much can be accomplished in addition to established programming. While it is commendable to adhere to a schedule, it is not always practical. Be flexible enough to make changes while keeping an eye on the ultimate goal: programming to support STEM/STEAM activities for young people to further their interest and skills.

It may be helpful to use a chart to monitor programming activities. The indicator of progress should be a step toward achieving the end goal. The progress category can be used to describe the intermediate steps to completing the activity or to record completion dates of the activities. In Figure 1.2, the activities listed are a continuum toward completion and evaluation of the goal. Figure 1.3 provides a blank template for your use.

Do not forget to celebrate successes. As progress is monitored, adjust implementation steps as needed. As you complete activities, the evaluation of the activity and the goal are important to refine goals and address any issues that come up in the process. Not making expected progress does not mean a goal is not appropriate. It does require further reflection and adaptation as needed.

Keep in mind the logistical issues that can crop up as you move forward. If you're in a school setting, identify colleagues who may be most receptive to collaborating on STEAM activities. Talk about ways in which you can support their work, and how they will benefit from your support. Their input will inform what you may want to plan for STEAM in collaborative lessons and in informal activities outside the classroom. For community resources, scheduling may be determined based on availability of local experts who are interested in working with students in these activities.

Activity	Indicator of Progress	Progress to Goal			
		1	2	3	4
Goal: RAISE AWARENESS					
Mtg with principal to discuss action plan	Notes from meeting		X		
Presentation at staff mtg re STEAM activities	Agenda, comments from teachers			X	
Goal: STEAM COLLABORATION					
Schedule mtg with science dept teachers to discuss collaboration	Minutes of meeting	X			
Schedule planning mtg for STEAM lesson with math & art teacher	Finished lesson plan	X			
Indicators: *1 = Identified* *2 = Planned* *3 = Implemented* *4 = Evaluated*					

Figure 1.2: Activity Tracking Sample

Activity	Indicator of Progress	Progress to Goal			
		1	2	3	4
Goal:					
Goal:					
Goal:					
Goal:					
Indicators: 1 = Identified 2 = Planned 3 = Implemented 4 = Evaluated					

Figure 1.3: Activity Tracking Chart

Enlisting Administrative Support

As with any new initiative, it is imperative to involve administrators and supervisors in the process. Talk with your administrators and supervisors to determine how best to approach this project. Without their support, it is unlikely that any programming will be successful. They are also not happy with surprises, so keep them informed. What happens if they are not supportive? The conversation may be turned to current efforts to implement STEM/STEAM in the school curriculum. In public library settings, you may have an opportunity to educate supervisors about the trends in education around this topic, and about trends with libraries as places for informal learning.

Make sure you know the administrators involved in this conversation, and keep in mind their operating style. For some, it may be prudent to preface the discussion with a disclaimer that the conversation is to discuss possibilities, with no need for an immediate answer. (In a busy setting, it is easy to say no to new initiatives.) Then a follow-up discussion can be held to get an okay to proceed, with any adjustments the administrators or supervisors see necessary.

Be prepared for questions about costs and procedures. Have a plan for an activity or two that could launch the program, with an outline of what might be needed. This may be where a "found materials" project could be most appealing, as a stepping-stone to further funding. If there are library funds available, outline how they will be used, breaking them out as a cost per student to show how economical it could be. The focus of the discussion is what this will do for students, for the young people being served. In a school setting, a focus on student achievement or growth is compelling. In a public library setting, the level of engagement and appeal to a specific patron group that become the taxpayers of the future may be productive.

What happens when there is opposition to the plan that you have presented? First, look at it as an opportunity to calmly discuss the administrators' concerns. Hopefully negative issues and roadblocks will be limited if there has been a continuing dialog. For instance, unexpected budget issues may arise that can influence funding, so other sources can be explored. Administrators may also have a broader view of issues at the building or district level that need to be considered. While any opposition can be distressing, do listen carefully so you can reflect and come up with possible solutions or responses to any questions that arise.

Enlisting Community Support

Either initially or when some survey results have been compiled from these documents, further questionnaires could be used for parents and for community groups. Building on findings from student responses, a cover letter could explain the rationale for STEAM programming. This information could be part of a letter used in requesting help with specific programs, or as a part of an advisory group.

Here is where documenting the programming can really benefit your efforts. Images from the sessions, and comments on exit tickets asking for what was learned or what was enjoyed the most can be used on social media and the library website. (For images, make sure you have proper permission forms for minors, signed by parent or guardian, stating that images can be used.) Any articles, blog posts, or other reports will help build support. Make sure the community knows, and share with administrators and supervisors.

GROWING THE PROGRAM

Successful program ideas grow incrementally. Keep in mind that time restraints and existing daily activities do not go away while you focus on STEAM activities and programs. Set realistic goals about how many sessions you can set up, how many teachers you can collaborate with within the existing schedule, and what resources can be assigned to STEAM within existing personnel and budget constraints. When using outside speakers and agencies, keep in mind that stipends or honorariums may be part of the picture for budgeting as well.

Rubrics can be used to help evaluate a STEM/STEAM program and related activities. There are rubrics available for full STEM curricula that could be modified to use in the library setting. For example, North Carolina developed the STEM Attribute Implementation Rubric for both high school and middle school. The Creative Commons license for this rubric means it can be used and modified for noncommercial educational purposes, with credit given to the developer. A sample from the rubric is shown in Figure 1.4. It shows the stages of adoption (early, developing, prepared, and model) for effective in- and out-of-school programs in a discussion of curriculum. See the reference list for the link to the full rubric.

Any rubric you use should be able to measure objective steps along a continuum. This example uses "early, developing, prepared, model" in both its high school and middle school versions (Friday Institute for Educational Innovation, 2013). Others may use different measures, like beginner-intermediate-expert, or started-in development-completed. Use the terms that work for your particular conditions. Then choose the categories you want to measure. This may include areas like project setup, facility readiness, participation levels, and so on. The RubiStar website has

KEY ELEMENTS	Early	Developing	Prepared	Model
(2) Curriculum: Connections to effective in- and out-of-school programs				
2.1 STEM Network	School/program is seeking to establish partnerships with other schools, communities, postsecondary institutions, and businesses to identify solutions for building a quality STEM school/program	School/program engages with other schools, communities, postsecondary institutions, and businesses to identify solutions for executing a quality STEM school/program	School/program has documented partnerships with other schools, communities, postsecondary institutions, and businesses for executing a quality STEM school/program	School/program has partnerships with other schools, communities, postsecondary institutions, and businesses to identify solutions for executing a quality STEM school/program; partnerships are purposeful, mutually beneficial, monitored, and evaluated
2.2 Students and STEM Professionals	Leaders are creating plans to provide opportunities for students to meet STEM professionals and/or to experience professional STEM work environments during and/or outside school	Direct experiences with STEM professionals, professional STEM work environments, and/or practical applications of STEM content during and/or outside school are available to students at least 2 times throughout the year	Direct experiences with STEM professionals, professional STEM work environments, and/or practical applications of STEM content during and/or outside school are available to students at least 4 times throughout the year	Direct experiences with STEM professionals, professional STEM work environments, and/or practical applications of STEM content during and/or outside school are available to students approximately monthly
2.3 Research & Development	On an annual basis school/program leaders and other STEM teachers share with each other research and information on best practices related to their STEM program goals	On a semiannual basis school/program leaders and other STEM teachers share with each other research and information on best practices related to their STEM program goals	On a quarterly basis school/program leaders and other STEM teachers share with each other research and information on best practices related to their STEM program goals	On a monthly basis school/program leaders and other STEM teachers share with each other research and information on best practices related to their STEM program goals

Figure 1.4: Programming Rubric Example

Source: Friday Institute for Educational Innovation. High School STEM Implementation Rubric. Raleigh: North Carolina State University. 2013. https://www.ncsmt.org/wp-content/uploads/2013/09/STEMAttributes Rubric_HIGH_v4_Aug2013_v1.pdf. Reprinted with permission.

From *Full STEAM Ahead: Science, Technology, Engineering, Art, and Mathematics in Library Programs and Collections* by Cherie P. Pandora and Kathy Fredrick. Santa Barbara, CA: Libraries Unlimited. Copyright © 2017.

a "rubric builder" that can be useful if you are new to rubric development. If you are working with a classroom teacher on a specific project, they may have already developed a rubric that you can use. The point of the rubric is to give a path not only for evaluating how far you have come, but what qualities a successful project or program will have on completion.

Programming Ideas

Your program can be a combination of informal learning opportunities paired with collaborative efforts with teachers in STEAM disciplines. The key is to build it carefully to coordinate these efforts with existing library programs and activities. As you plan programs, whatever the scope, keep in mind elements that promote hands-on learning, such as:

- Identify concept(s) to be learned and making this clear to attendees.
- Provide open-ended situations and the materials needed for problem-solving.
- Allow for trial and error as an important part of the process.
- Give opportunities to observe and test products, making modifications as needed.
- Demonstrate what's being done whenever possible—model the behavior expected.
- Encourage peer-to-peer learning and group activities.
- Talk through the creation process, letting students explain what they did.
- Provide library resources to support any activity.

Here is a grab bag of starter ideas to use in planning activities in your setting that can become part of the library program. Some are one-off ideas; others can be the basis of a series of sessions or a unit of study. Use these ideas to spur or supplement collaborative efforts with teachers, or be stand-alone options for informal learning for young people during and after-school.

Set up a station with a weekly challenge for students to attempt. This can be as simple as creating a structure with a deck of cards; building a Lego structure to hold a certain amount of weight; designing a structure on paper; or completing an electronic circuit experiment.

Host a robot competition à la the television program, *BattleBots*. Rather than focusing on destruction, the competition could hinge on how much weight a robot could carry; a specific programming challenge for the robot to meet; or a specific set of quests on a circuit to be completed by the robot.

Students develop measurement tools, using the design process to develop a new ruler, multidimensional measurer, etc. Reports are written as news items on this project, explaining the project and solution.

Develop a speaker series focused on careers in STEM fields. Encourage speakers to bring hands-on materials or items to share with the students involved in the program. In a school setting, this could be a collaborative lesson with science and technology teachers.

Provide an area in the library for students to tinker. This are can include construction supplies, measurement tools, activity kits, and resource books. On the library's website, provide links to sites that will pose challenges, encourage creativity, and give an outlet for problem-solving. (See Chapter 9, Makerspaces, of this book for details.)

Set up a display area for young people to display their creations from makerspaces, class projects, and other sources. This passive public display will encourage others and spark interest.

Extend the reach of a science-based activity to involve the arts, whether it is elements from an art class or from performing arts. An activity around designing flying machines can extend to playing music that is evocative of flight, or choreographing a dance sequence to illuminate the experience of flying.

Offer parent sessions to expose them to STEAM activities. In a school setting, this could be set up at a curriculum night or open house. In a public library setting, it could be part of adult programming. These sessions could be open to the community, including business associations or civic groups.

Use science fiction as a jumping-off point for coordinated activities around STEM themes and subject standards. Artists can recreate a world in images, mixed media, or models, while describing what characteristics the planet has that could impact life on the planet. Create the life-form that could exist on the designed planet, providing explanations about what scientific principles affect the life-form.

Develop a project based on an identified community problem in conjunction with science teachers. Students can brainstorm issues with guidance from adults. Schedule meetings or phone conversations with local contacts and experts in the field chosen. Students can build a plan for dealing with the issue, with end products including video, posters, presentations, and speeches that are presented in a public exhibition involving local participants and experts contacted. This could involve energy saving, recycling, gardening, connecting farms with town/city for produce distribution, and so on.

Host a puzzle day, where young people are invited to solve or create puzzles, either individually or in groups. Small rewards could be given for those who complete puzzles the quickest, for those who design a puzzle by age category, and so on. Encourage participants to use mathematics concepts to design the puzzles.

Have teams use a design process as they develop a new app or device to build a model for a new windmill. Build a model of the windmill. If there is a 3-D printer available, each team can produce their design.

Set up a recycling challenge first to get reusable materials to stock your materials for STEAM activities. Students can set up the challenge, using spreadsheets to track materials and calculate the top providers, whether in groups or individuals. Use a recycling theme to begin a project where students use materials to create a piece of artwork that meets certain mathematical principles (geometric shapes, exponential numbering, etc.). It could also be based on reading about recycling programs in journals or on a field trip to a local recycling center.

Involve young people in fundraising for an identified cause. Arts, mathematics, science, and technology can be involved depending on the cause. Students can develop a plan of action, determining how they will measure the success of the venture. Tracking expenditures and revenue can involve math skills and use of a spreadsheet (technology). Getting the word out about the event or activity can involve communications via video, fliers, or announcements (art and technology). Organizational skills are required to plan, implement, and conduct the event. Follow up with reflection time and review of what could be done differently in a future event or activity.

Host an exhibition with a STEAM connection. The event can focus on gaming, video products, or robotics, for instance. This could be a culminating event to a workshop or series of sessions around these areas. Young people can be invited to bring their creations and share their work with others in a tabletop discussion format or a science fair setup where participants, friends, and family can come to learn more.

In a school setting, it can be rewarding to set up activities to support STEAM. In public libraries, it can be a vibrant program with lots of buy-in from young adults. In either setting, the level of involvement can grow by involving young people in the growth of the program. Make it known that you are looking to provide materials for them to create. Let them come up with ideas—a new twist on the suggestion box. Make sure to listen as they talk about what works for them. As often as possible, act on their ideas to build engagement and enthusiasm for the program.

Collaborating with STEAM Teachers and Fellow Librarians

As ideas bubble up about STEAM activities and sessions, they can be part of library activities. In a school setting, the deepest curriculum connections grow from collaborating with other educators from the initial planning stages to the evaluation of activities and units of study. Once you have established relationships with those who are interested in collaborating, be proactive. Make sure to get in on the initial discussions of what is to be accomplished for a given lesson or unit. With the time constraints teachers are under, start with one project and find a receptive teacher (or vice versa). Get in on planning level. Once a teacher comes in with a prepared lesson, your opportunity to share library standards that apply may have passed.

Keep in mind that guided inquiry, which is a hallmark of library instruction, is reflected in science, arts, and engineering standards calling for a process for action, a scientific process, and systems planning. Exposing students to this across curricula enriches their experience and helps them see connections across disciplines. STEAM focuses on hands-on learning, a natural pairing with project-based learning. Match standard in each discipline with a project that engages students and gives them opportunities for active learning. As appropriate, bring together teachers from multiple disciplines for discussion, so that connections can be made between the arts and STEM teachers. Frame discussions in terms of essential questions, hypotheses, process development, inquiry, or any key words that will help those involved look at connections between disciplines.

In any design or scientific process, there is a research element; librarians can plug in most easily to this area. Remember that in our role as a curriculum consultant, we are not limited to the research paper or project. So plug in to research connections, but don't limit yourself to this. Encourage teachers to differentiate project ideas in their planning to give options for all sorts of learners. Offer space, materials, and a place to do a public display of students' finished products. There are two motives in having exhibits of student work. First they affirm the work and creativity of young people. Second, they provide a focal point to spread the word about collaborative work, opening doors to further collaboration.

Usually the school librarian is a solo position of its type in the building, and sometimes in the district. In public libraries, the youth services position may also be limited to a small number of people. Extend collaboration beyond the walls of each institution to collaborate with fellow librarians. It may be hard to find time, but with social media and communication tools, planning does not have to be a physical face-to-face connection. At the least, you can share what each institution is offering to young people. At the most, you can feed off the ideas of the other, and develop events from one site to another. Again, time limitations can mean starting small: share information about current activities or plan one specific event that benefits both the library and the classroom activities. There is strength in numbers and in the ideas of like-minded professionals.

Resources for Programming

Alessio, Amy J., Katie LaMantia, and Emily Vinci. *A Year of Programs for Millennials and More*. Chicago: American Library Association, 2015.

Braun, Linda W. "Want to Start a STEM Program? Assess Your Community Needs First." *School Library Journal Extra Helping eNewsletter*. February 29, 2016. http://www.slj.com/2016/02/teens-ya/want-to-start-a-stem-program-assess-your-community-needs-first

C-STEM Curriculum Modules. http://www.cstem.org/curriculummodules/ C-STEM was developed by Dr. Robin Flowers particularly to involve minorities in STEM activities to lessen the achievement gap. As a part of the annual C-STEM competition, enrolled educators receive curriculum materials to use with students in developing projects.

Jolly, Anne. "How to Design a Successful STEM Lesson." CTQ Collaboratory. Blog. September 28, 2016. http://www.edweek.org/tm/articles/2016/09/23/how-to-design-a-successful-stem-lesson.html?r=115092607

Koestler, Amy. "30 Days of Teen Programming: Engaging All Teens in STEAM: Thinking Diversely, with a Program." YALSA Blog. April 10, 2015. http://yalsa.ala.org/blog/2015/04/10/30-days-of-teen-programming-engaging-all-teens-in-steam-thinking-diversely-with-a-program-plan
One of the number of posts on the YALSA blog involving STEM and STEAM activities and discussion, including programming ideas and strategies for implementation.

littleBits Education. "How to Start a STEAM Program in Your School." https://s3.amazonaws.com/littleBits_pdfs/EDU-STEAMGuide-V1-5.pdf
This vendor publication has practical steps for building a STEAM program, including examples from schools and districts, research connections and the rationale for STEAM in schools.

LRC STEM Curriculum Materials. University of Wyoming Libraries. http://libguides.uwyo.edu/c.php?g=236800&p=1573160
This LibGuide features the collection of STEM materials in the Learning Resource Center highlighting collections and providing links to STEM activities.

RubiStar. http://rubistar.4teachers.org/index.php
Among the many tools for teachers on the 4teachers.org site, RubiStar is a standout. With a free account you can use their templates to create, save, and share rubrics for a variety of projects and purposes.

SciGirls. Connecticut Public TV. http://cptv.pbslearningmedia.org/collection/
scigirls/
These half-hour programs focus on involving girls in STEM activities
through seven-program modules including "Engineer and Invent," "Technology," and "Citizen Science," among others. Support materials are also
provided.

STEM Ready America. http://stemreadyamerica.org
This site brings together information in support of after-school and summer STEM programs, with a compendium of articles about success stories
covering programs, policies, and practices.

"Workshop: Interdisciplinary Programming in Your Classroom." WNET Educational Broadcasting Corporation, 2004. http://www.thirteen.org/edon
line/concept2class/interdisciplinary/index.html
Led by noted educational consultant Heidi Hayes Jacobs, this workshop
covers all aspects of interdisciplinary programming, from explanation to
exploration to demonstration to implementation. While focused on the
classroom, it is valuable to broaden understanding of how best to bring
multiple disciplines together to strengthen learning options for students.

YALSA. "Issue Brief #3: Libraries Help Teens Build STEM Skills." American
Library Association. http://www.ala.org/yalsa/sites/ala.org.yalsa/files/
content/IssueBrief_STEM.pdf

YALSA STEM Resources Task Force. "STEAM* Programming Toolkit." Young
Adult Library Services Association, American Library Association, 2014.
http://www.ala.org/yalsa/sites/ala.org.yalsa/files/content/2016%20STEAM
%20TOOLKIT.pdf

YALSA STEM Resources Task Force. "Stem Resources." http://wikis.ala.org/yalsa/
index.php/STEM_Resources

REFERENCES

Friday Institute for Educational Innovation. *High School STEM Implementation Rubric*. Raleigh: North Carolina State University. 2013. https://
www.ncsmt.org/wp-content/uploads/2013/09/STEMAttributes
Rubric_HIGH_v4_Aug2013_v1.pdf

Friday Institute for Educational Innovation. *Middle School STEM Implementation Rubric*. Raleigh: North Carolina State University, 2013.
https://www.ncsmt.org/wp-content/uploads/2013/09/STEM
AttributesRubric_MIDDLE_v4_Aug2013_v2.pdf

FURTHER READING

Batykefer, Erinn, Laura Damon-Moore, and Christina Jones. "The Library as Incu-
bator Project." Blog. http://www.libraryasincubatorproject.org Started as
an advocacy site for library/art connections, this site includes interdisciplinary ideas for STEAM projects.

Collins, Cathy. "STEM and the School Library: A Marriage That Makes Sense." *Knowledge Quest*. Blog. April 29, 2016. http://knowledgequest.aasl.org/stem-school-library-marriage-makes-sense

Gingrich, Christie. "String Art, Steam and Teens." *A Geek in Librarian's Clothing*. Blog. April 11, 2016. https://www.geekinlibrariansclothing.com/2016/04/11/string-art

Hinton, Marva. "Study Links After-School Programs to Improved STEM Knowledge." *Education Week*. Blog. March 1, 2017. http://blogs.edweek.org/edweek/time_and_learning/2017/03/new_study_examines_link_between_after-school_programs_stem_knowledge.html

Idaho Commission for Libraries. STEAM Resources. http://libraries.idaho.gov/page/steam-resources-tweens-and-teens. The website section shares program ideas.

Mies, Ginny. "Think Outside the Box When It Comes to Teen STE(A)M Programs." *Tech Soup for Libraries*. June 11, 2015. http://www.techsoupforlibraries.org/blog/think-outside-the-box-when-it-comes-to-teen-steam-programs

Schadt, Erin M. "How to Create a Robust STEM Program." WebJunction. May 3, 2016. http://www.webjunction.org/news/webjunction/how-to-create-a-robust-stem-library-program.html

Stephan, Michelle. "Off to the Duck Races: Planning for Inquiry in STEM." *Educational Leadership* 74, no. 2 (October 2016). http://www.ascd.org/publications/educational-leadership/oct16/vol74/num02/Off-to-the-Duck-Races@-Planning-for-Inquiry-in-STEM.aspx

Techman, Melissa. "New Ideas for Better STEAM Programs." *School Library Journal Online*, October 19, 2015. http://www.slj.com/2015/10/programs/new-ideas-for-better-steam-programs/#_

2

◇ ◇ ◇

FUNDING

FUNDING YOUR PROJECT

One of the difficulties in creating new programming is determining what you need and how to gather the funding needed. Your first step after you've set your goals is to research the items that you want and determine how much they will cost. Create an action plan with a timeline and a tentative budget. See Figures 2.1 Budget Template and 11.1 Action Plan for suggestions.

Meet with your direct supervisor, manager, or headmaster to get his or her thoughts and tweak the plan as needed. While you don't need to have a multipage report ready you do need to be prepared. To convince him or her of the merits of your project, come armed with a list of benefits derived from the project, specifically how it will benefit your teens, your staff, and your community. Relate your project to departmental and library goals. If you're in a school library identify the standards that will be met. The more information you have gathered the better. Limit your action plan and budget to one-page summaries each and leave copies with your manager.

Anticipate the questions that you'll be asked and prepare to justify your plan with rationale for expenses and purchases. Do not be discouraged if you meet resistance. Many managers are inclined to say "No" at first. If you receive a negative response politely ask for specifics. Ask if your

Category	Quantity	Item	Unit Cost	Total Cost
Crafts Rationale:				
Equipment Rationale:				
Furniture Rationale:				
Kits Rationale:				
Supplies Rationale:				
Software Rationale:				
Tools Rationale:				
Totals				$0.00

Figure 2.1: Budget Template

supervisor objected not to the premise of the project but could not support it monetarily. Thank the person for the time to listen. Don't argue, but don't despair.

Reflect on what you've learned and your rationale. The rejection may help you to build a stronger case in the future. Rework your numbers and revise your pages. Confer with your library colleagues and any partners or teachers who are part of the proposal. Set it aside for a few weeks before you revisit it. The short break will give you time to mull over alternative scenarios and help you to build a more persuasive argument.

Locate the policy manuals that govern your library. Once you have received approval from your direct manager or principal, make an appointment with the financial officer, CFO, or treasurer. Your purpose is twofold: first, to convince this person that the project is worthwhile and will reflect well on the library, the school, and the community. Second, you need the information that goes beyond the manuals, also known as past practice. You need to know the informal policies that need to be followed before requesting funding or accepting donations from outside agencies. Some boards require prior approval and you need to know these restrictions before you approach anyone outside the library. You don't want to be embarrassed or embarrass your manager by circumventing the system already in place.

Start In-House

Start in-house. Use your action plan (see Figure 1.1), timeline, and a list of items that you want to purchase to determine how to finance your plans. When you share your action plan with your manager identify the amount of your library budget you can expend for relevant supplies, equipment, databases, books, and electronics without compromising current programming. Let the manager know the specific amount you're requesting from his or her budget and what items you want to purchase with the money (Pandora and Hayman, 2013, 151). Try to utilize a combination of funding from your budget and your manager's budget with additional assistance from other departments, that is, children's or tech support in the public library, or science, math, art, or music departments in a school setting. The culture of your library will determine how you approach other departments. If you have a good working relationship with a department and collaborate often, you may be comfortable asking that supervisor or department head to share in the project costs. In other libraries, however, you aren't free to solicit help from other departments without prior approval from the boss as it can be difficult to separate and share funds (Squires, 2016, 137–38).

Monetary Donations

Donations are one of the most cost-effective means of starting your programs (Rendina, 2016). Be sure that you know the policy that has to be followed about donations. If your school or library has a foundation or Friends group, they may be helpful in locating funding. By this time you should know whether you need the approval from the trustees or school board before accepting donations.

Start an Online Library Fund

In our experience: Shaker Schools Foundation set up a special "Library Fund" so that people could donate just to the libraries.

Is there already a donate button on your library's website? If so, ask if donors can earmark their funds for your project. Add a line so that individual donors can list their company names if their companies will match donations (Rossman, 2016, 148). Corporate partners may be willing to provide event sponsorship or become sustaining donors offering money over a specified period of time (Rossman, 2016, 157).

Or try an old school idea made new again. Elementary school libraries have traditionally funded new books through a Birthday Book Club; public libraries could do so as well. Parents send in a check of a predetermined amount and the birthday child selects a new book. A book plate with the child's name is placed in the book and the child is allowed to be the first borrower of "his/her" book. If your library policies already allow this, you can institute a similar program for end of the year celebrations, noting that there are now multiple promotion ceremonies—kindergarten, 6th, and 8th grades as well as the traditional graduation after senior year.

Legacy gifts can be large sums, such as a memorial naming of the library, or smaller sums, such as books, or equipment. Naming rights donations are often given in memory of someone and are usually reserved for large projects, such as naming a particular room. If you are renovating the reference area, creating a makerspace area, a teen room, or a computer lab you may be able to secure funding by selling the naming rights to the new facility. Again, check your policies very carefully. This is one case where it is wiser to secure permissions first; be sure to speak with your treasurer or fiscal officer before soliciting these donations!

Donation of Goods

Do the rules differ for monetary donations versus donations of goods? Don't take donations unless, and until, there are policies in place with parameters of what will be accepted giving you the right to dispose of those items that do not meet your needs. Numerous sample policies can be found on the library association web pages. The ALA Office for Intellectual Freedom "Workbook for Selection Policy Writing" guides you through the process of writing your own selection policy if you do not have one already in place (2017, http://www.ala.org/Template.cfm? Section=dealing&Template/ContenManagement/ContentDisplay.cfm& ContentID=11173).

The CREW Weeding Manual from the Texas State Library and Archives Commission (2012, 27, www.tsl.state.tx.us/ld/pubs/crew/index.html) has a section on donations that can be revised and added to your own policy. The selection policy should regulate discarding print materials or electronics deemed too old, too moldy, too corroded, or too out-of-date to be useful. All libraries, both public and school, must have these policies in place BEFORE requesting donations. IF not already in place, then create a policy that contains the following crucial elements:

- The right to evaluate all items for use
- The right to discard or donate items

Use the *Texas Sample Donation/Weeding Policies* as a guideline (https://www.tsl.texas.gov/sites/default/files/public/tslac/ld/ld/pubs/crew/crewmethod12.pdf).

You don't want to face 40 years' worth of *National Geographic* magazines that come in from every family that cleans out an elderly parent's basement. Analyze each donation to determine its worth for your library and your project. If the item does not meet your needs, or you already have all that you need, the policy provides written permission to reduce, recycle, or discard.

Decide, then, if the donated items are of value to another agency. Send quality, duplicate print materials to the public library for the Friends of the Library book sales. Sending children's books and fiction to women's and homeless shelters is another avenue for items you don't need *if, and only if*, they are quality items in good condition. Many nonprofits, including schools and cities, have bins for the purpose of recycling paper. If you deem print materials of no use to anyone, then cancel them, cut off the cover of the books, and send the pages to recycling. (Granted, this is a hard thing for librarians to do.) Local zoos shred old phone books and paper for bedding for animals and may welcome your donations. Electronic items should be analyzed by your tech department; if found unusable as a whole then the parts can be cannibalized for use in repairs.

ALTERNATE FUNDING SOURCES

Once you are familiar with the policies regarding donations, and have determined that monies are not available from other in-house sources, search for alternate means of funding your programs. These alternates may include the following:

1. Associated funding sources, for example, the parent/teacher associations or organizations (PTA/PTO), Friends of the Library
2. Trustee's special funds or the school's educational foundations
3. Local service organizations
4. Grants
5. Social media, for example, Crowdfunding

Associated Funding Sources

While departmental and manager's budgets may be depleted, there are some organizations associated with the library that can provide monetary relief. The PTA/PTO support schools through volunteer work and with donations. Ask your principal if they make donations directly to the principal's account or if they allow requests from faculty. Ask if there is a particular time of year, for example, July or early fall, that is best for requesting funding. Note that even if their budget is depleted these parents can supply you with volunteers who can save you staff time. In the public library the Friends of the Library is the group that runs the fundraiser book sales and provides volunteers and funding. Ask your director the same types of questions listed for the PTA/PTO.

Trustees' Special Funds and Educational Foundations

If the board of trustees has a fund for special projects you cannot usually request funding directly from them. Channels have to be followed and requests should be initiated by your director rather than a staff member. The action plan, budget, rationale, and other documentation that you have compiled are especially valuable here as someone else will be making the case for your project.

Educational foundations are a great bonus for school libraries. Donations are tax deductible so there is an incentive for community members to donate (Pandora and Hayman, 2013, 154). They provide grant opportunities for faculty members and they love to see that the faculty have partnered with other institutions. This is a perfect fit for the school librarian who works with an entire school, or multiple schools, with numerous teachers, and, we hope, with the teen librarian at the public library. Their applications are shorter than those for federal grants.

In Our Experience

Ed foundations love libraries! Our educational foundation loved our library. They funded numerous programs and supplied matching funds for LSTA grants that helped us to build library computer labs.

Service Organizations

Service organizations are great source of local funding. These organizations commit to funding community nonprofits including schools and libraries that benefit young people. Review the businesses that purchased ads in various school publications, such as sports, band, theater, dance programs, school yearbooks, newspapers, and literary magazines. As advisor to our yearbook staff, I was surprised to learn that some of the most generous businesses in town were the insurance agencies and funeral homes. Ads in student publications and programs were a great way for them to get their businesses noticed. Chambers of Commerce may not supply monies directly but will be able to help you to identify businesses that can support your endeavors and may allow you to present your STEAM project proposals at a meeting. As an organization of local businesses they can point you to experts in the areas of banking, finance, engineering, science, photography, music, and art who may be willing to spend time with your patrons or serve as advisors for you, or mentors for your teens.

Many of these organizations do not require a formal grant application but will entertain a proposal for funding of specific projects. Welcome all requests for presentations explaining your project and have some young adults present with you. You never know who will be interested or who will have a connection to a partner you need. Plan to make presentations after the project to any groups providing funding. Be sure to thank them for their support in person as well as with a written thank you note. Not only are written notes good PR but they can be read aloud at the group's meeting and included in the formal minutes. Organizations like to fund nonprofits that have a proven track record and that are grateful for their support.

Grants

If you've exhausted the limits of your budget and need another source, look for grant sources. Grants help you to afford items that are outside the scope of your ordinary budget and are especially helpful when you have a specific project in mind (Rossman, 2016, 78). Local funding sources should be one of your first items to research. Look at the listing in Figure 2.2, Grant Sources.

While you will not find all of the organizations listed on the chart in your community, you will find similar philanthropic groups in your area. Local businesses and agencies have a stake in their communities and are more likely to require short forms, letters, or conversations rather than multi-page applications. Schools often have educational foundations that fund teacher requests. Since most prefer that the school partner with outside agencies this is a great project for a school-public library collaboration.

Use this checklist to investigate potential sponsors in your area:

- ☐ Educational foundations
- ☐ Kiwanis
- ☐ Rotary
- ☐ Eagles
- ☐ Lions
- ☐ Moose
- ☐ Elks
- ☐ Other service organizations
- ☐ Veterans of Foreign Wars
- ☐ American Legion
- ☐ Funeral homes
- ☐ Insurance companies
- ☐ Chamber of Commerce
- ☐ Garden clubs
- ☐ Crowdfunding, e.g., Donors Choice, Kickstarter . . .
- ☐ Women's groups
- ☐ Grants.gov
- ☐ Library Service and Technology Act grants (check your state library)

Figure 2.2: Grant Sources

Goal	Promote critical thinking using STEAM activities Purchase robotics kits for Year Two
Project	Create a Lego wall advertising new makerspace
Objectives	Meet NGSS and state standards for Science and Math
Methods	1. Create a makerspace area with existing furniture 2. Pursue donations of craft and building blocks 3. Locate an engineer to speak to Teen Advisory Board 4. Promote building activities via photos and intranet PR 5. Identify teachers to use the makerspace
Partners	Public library Chamber of Commerce speaker's bureau
Staff	School librarian Teen librarian from public library Parent volunteer
Evaluation	User satisfaction surveys Desire for continued programming Teen enthusiasm (focus groups and observation) Teen participation
Sustainability	Purchase crafting supplies from consumable budget Solicit donations for circuit boards
Results and unexpected consequences	To be completed during and after the project
Conclusion	To be completed during and after the project

Figure 2.3: Grant Format Sample

Whichever source you choose to use your grant may look something like that in Figure 2.3. A blank copy is available in Figure 2.4.

If you have a large project or need additional monies, start the search for other grants, those that are subject-specific, for example, the sciences, or that are designated for a particular group of individuals, say for females or minority students for instance. National companies such as Best Buy, Dollar General, Lowe's, and others have grant applications and explanations online (Rendina, 2014). Other companies provide monies directly to the principal of each school. Target and Walmart are two companies that give money to the schools in the cities where they are located. Research the big box stores and those with national franchises to see if this is a possibility for your project. There are also online grant resources such as Grants.gov that can be checked. Call or e-mail your state library staff as well. Library Service and Technology Act (LSTA) grants are administered through each

Goal	
Project	
Objectives	
Methods	1. 2. 3. 4. 5.
Partners	
Staff	
Evaluation	
Sustainability	
Results and unexpected consequences	
Conclusion	

Figure 2.4: Grant Format Template

state with monies from the Institute of Museum and Library Services (IMLS) and each state has a unique way to serve its library patrons.

If your library has the IRS designation of 501(c) (3) then you may be eligible for federal grants as well. Matching grants, especially federal grants, require that you provide a percentage of the budget before they will match it. Most often you will need to supply 25 percent of the total project cost while they match, and supply the remaining 75 percent. Funding agencies want a sign of your commitment to the project and provide this "seed" money with the expectation that you will sustain the project after the grant period is over (Rossman, 2016, 80).

Whether the grant you seek is local or national, identify your needs and describe the project as succinctly as possible. Identify your goals and objectives, your methods, and which staff will take part. Determine which evaluation methods you will use. Finally, consider how you will continue to fund the project in the future as funding agencies like to see that you can sustain the project after their money is expended. Create a sample budget that you can include and any sample evaluation tools that you plan to utilize. For a listing of additional grant sources for schools and libraries turn to the Resources section at the end of this chapter.

Social Media Sources

Again be sure to refer to your policies. Some libraries, particularly in schools, have very strict policies on what items can and cannot be placed on social media websites. In addition to donations and grantwriting, social media sites are a means of attaining funding to expand your budget. Sites such as Kickstarter, Crowdfunding, Donors Choose, and other donation sites allow you to present your requests to both local and international individuals for help in achieving your goals. Some have a time limit as to how long your project will be posted. Be aware that some have an all-or-nothing policy; if your project is not fully funded by the agreed upon date then you will receive none of the funding (Rossman, 2016, 68). The reasons these are popular are twofold—the information you provide is short and is easy to share via social media channels and the library's website and Facebook pages.

COMMUNICATION

Be sure to give progress reports to your funding agencies and your stakeholders. They need to know how their money is spent and that it is spent wisely. Let them know how you are doing and how your teens are reacting. Ask how often and what forms of communication they would like. Funding agencies may prefer e-mail updates while your manager may want an initial meeting with one page summaries at the midpoint and end of the project. For more on reporting successes, failures, and unexpected consequences, see Chapter 11.

RESOURCES

Best Buy Foundation. https://corporate.bestbuy.com/best-buy-foundation-nati onal-partnership-request-for-proposals-2015-2. Provides grants to schools.
Dollar General Literacy Foundation. http://www2.dollargeneral.com/dgliteracy/ Pages/grant_programs.aspx#ylg. Includes grants for adult literacy, youth literacy, summer reading, family literacy, and Beyond Words: The Dollar

General School Library Relief Fund for schools that have suffered a major disaster.

Grants. https://www.grants.gov. Means to search for grants from the federal government. Links to eligibility criteria, granting organizations, and forms.

Grants for Libraries. http://librarygrants.blogspot.com. Listing of grants just for libraries.

Grants for Teachers. http://www.grantsforteachers.com/foundation-grants/ Dollar_General_Literacy_Foundation_Youth_Literacy_Grants/grantde tails_137.aspx. Search teacher grants by subject, state, or grade band. One of the topics is STEM.

Lowe's Toolbox for Education Grant. http://toolboxforeducation.com. Donations to schools and nonprofits.

National Leadership Grants for Libraries. https://www.imls.gov/nofo/national-leadership-grants-libraries-fy17-2-notice-funding-opportunity. Administered through the Institute of Museum and Library Services (IMLS).

REFERENCES

ALA, Office for Intellectual Freedom. "Workbook for Selection Policy Writing." 2017. http://www.ala.org/Template.cfm?Section=dealing& Template/ContenManagement/ContentDisplay.cfm&ContentID= 11173

Larson, Jeannette. CREW: A Weeding Manual for Modern Libraries: Texas State Library and Archives Commission, Austin, Texas, 2012. www .tsl.state.tx.us/ld/pubs/crew/index.html

Pandora, Cherie, and Stacey Hayman. *Better Serving Teens through School Library-Public Library Collaboration*. Santa Barbara, CA: ABC-CLIO, 2013.

Rendina, Diana. "How to Start a Makerspace When You're Broke." Knowledge Quest Online, February 22, 2016.

Rendina, Diana. Lowe's Toolbox for Education Grant. "Library Makeover Preview." Blog. May 23, 2014. Toolboxforeducation.com

Rossman, Edmund A. *40+ New Revenue Sources for Libraries & NonProfits*. Chicago: ALA, 2016.

Squires, Tasha. *Library Partnerships: Making Connections between School and Public Libraries*. Medford, NJ: Information Today, 2009.

FURTHER READING

Fink, Jennifer. "Crowdfunding the Classroom." *District Administration Online*. September 2016.

Gerding, S., and P. MacKellar. "Grants for Libraries: A How-to-Do-It Manual and CD-ROM." New York: ALA, Neal-Schuman, 2006.

McLaughlin, Molly K. "The Secret to Crowdfunding Success for Inventors and Backers." *PC Magazine Digital Edition*. October 2016. http://web.a.ebscohost .com/ehost/pdfviewer/pdfviewer?sid=31af927c-4821-4607-93eb-427a714 28be0%40sessionmgr4010&vid=1&hid=4207

Pledger, Marcia. "Manufacturing Contest [M]SPIRE Seeks Online Applicants for Grant Money." *The Plain Dealer* (Cleveland, OH), August 26, 2016. http://www.cleveland.com/business/index.ssf/2016/08/manufacturing_contest_mspire_s.html

Rendina, Diana. "STEM Maker List Grant." Blog. July 30, 2014. http://renovatedlearning.com

3

◇ ◇ ◇

OUTREACH AND COLLABORATION

OUTSIDE THE LIBRARY'S WALLS

Librarians collaborate daily with experts in other departments. Teen librarians work with reference librarians, children's librarians, and managers. School librarians work with teachers in all disciplines and many parental groups. All librarians work with their directors, principals, headmasters, and information technology staff. They usually have contact as well with the trustees and school board members who oversee the operation of their libraries. Librarians may informally work with each other through committee or community work. While both school and public libraries have lost staff in recent years due to budget cuts, we believe that collaboration between school and public librarians is the first step in building partnerships with outside agencies. We are more similar than different (Flowers, 1998, 105). We love our patrons. We enjoy helping them to find what they need. We encourage them. We work long hours to prepare programs and/ or lesson plans, reports, and presentations for those we serve. We know that our careers will not make us rich; we gain our satisfaction not from our salaries but from the satisfaction of helping every reader to find his or her resource—an update of Ranganathan's Second and Fourth Rules of Librarianship, "Every Reader His Book" And "Save the Time of the Reader" (Wallace, 2001, 2).

Outside the Library's Walls

"Some of the most important library work I do is outside the library's walls."
Abby Johnson, "Reach Out through Outreach." *American Libraries* 45,
no. 11/12 (November, December 2011):48

To truly serve our patrons we need to venture outside the comfort of our four walls (Johnson, 2011, 48). She believes that librarians need to be out making contact with the members of the community. The work you do outside your workplace will help you better understand your community and make others aware of the important mission you serve for young people.

Cooperation

It is likely that public libraries and school libraries have cooperated in the past, sharing information about upcoming programs and promoting each other's activities, such as library card sign-ups and summer reading. Cooperation is a first step but doesn't necessarily involve face-to-face contact. It consists of sharing materials electronically via e-mail, chat, social media, or website postings. Librarians share newsletters as well as teacher assignments; they send each other calendars and posters to promote special events. Collaboration takes more time but the value is immeasurable; the rewards are immense for the librarian as well as for patrons.

From Cooperation to Collaboration

To truly benefit your patrons, particularly in STEAM activities, you need to advance beyond the step of cooperation to collaboration.

Collaboration

It will take a little more time in the beginning, but if you collaborate you will both gain so much from the partnership. More importantly, you will be adding to the services that you give to your common patrons. We have to remember that sometimes our work is done best when we leave our own library and reach our patrons elsewhere (Braun, 2015, 58). This may take us to another library, a university, or a social service agency.

Here is a common situation faced by public librarians. Without communication of some kind, students bringing assignments to the public library may encounter unintended problems. Perhaps a particular teacher assignment is difficult in part because students don't fully explain the assignment or don't have it with them when they visit the library. If you routinely send assignments to the public library they can be forewarned and know which of their resources can help your students during the evenings and weekends when the school library is closed.

If you have become colleagues your teen or reference librarian will be more willing to tell you the problems and issues involved. Collaboration is much easier to foster thanks to social media tools that are available. Whether you and your potential partner prefer chat, blogs, another form of messaging, or video conferencing you have many opportunities to "meet" online (Squires, 2009, 111). While meeting for coffee can be hard to fit into a busy work week, we urge you to make the time. Meeting face-to-face with your counterpart provides you with another layer of partnership that can't be duplicated. If your libraries already have meetings set in place count yourself as one of the lucky ones. While cooperation involves working in a parallel fashion—working separately but sharing—collaboration requires working together toward a common goal. Collaboration involves a partnership with give and take, compromise, and flexibility. As members of the same community we share common patrons, parents, and senior citizens and are often asking that public to support us with their hard-earned tax dollars. It is helpful to have an ally who understands what we do and why we are so committed to our patrons.

One of the many advantages to this collaboration is the sharing of the workload on programming. You now have two collections and two budgets that can be tapped as you plan a joint program. School library budgets often don't allow monies to be spent on food or beverages. For those you have to talk with your principal or headmaster. Public libraries may be able to supply this item while the school uses a part of its budget for other items such as consumables.

The reasons for working with a partner go beyond saving money and working together. In terms of your patrons your partnerships will provide better customer service, serve needs that have as yet been unmet, and solve problems (Todaro, 2005–6, 143). Your partnership will help with collection development. As you create bibliographies from both collections you can avoid duplication and better refer your patrons to the other institution. When planning programs you now have a partner with whom to divide the research work load and duties.

EVENTS

Teen librarians can attend New Teachers meetings (held prior to the start of the school year) or Meet the Teacher night to greet the new teachers. School librarians often host Open House or Curriculum Nights with snacks and coffee to draw parents in so they can meet parents and distribute flyers with database passwords, hours, and homework help tips. Why not ask the teen librarians to join you with a table to promote their own homework help? Likewise, school librarians should attend special celebrations at the public library to learn the staff members there. If programs are held of mutual interest, perhaps the school librarians can showcase how they support teens in their institution. While you might not need permission to visit the other library it is wise to let your administrators know that you are doing so in order to provide better services for your teens (Squires, 2009, 85).

Invite the Community

Your first foray into the community at large might be as a partner librarian. Continue by inviting community members into your world, whether via publications, online resources, or visits to the library. In all community public relations materials, be sure to spell out that BOTH libraries are necessary for the community (Flowers, 1998, 106). Don't stop at inviting the public; be sure to extend your special events invitations to legislators, the media, and to your directors and superintendent (Flowers, 1998, 108). Invite influential parents and community members especially those on the school board or board of trustees. Figure 3.1 provides additional ideas.

Invite the mayor or any powerful politician who lives in your town, your county, your state, or your province. Politicians are interested in the positive workings of schools in their districts so invite them to events that involve your teen patrons in something active rather than passive. Note that they do like to have their photo taken with young people, however, due to their voting schedule in city hall, the local statehouse, in Washington, D.C., or Ottawa you will most likely have a visit from a staff member from their local office. Do not despair as this person will have the ear of your politician.

Invite the Media

Media presence is welcome, as this is how you get noticed. Realize that most often it is the print media stringer from the local weekly paper who will attend. It is very difficult to get the reporters from the local television or radio stations to attend unless your managers have a contact

Partner Agencies	Notes
4 – H Clubs	Biology, Zoology
Aquariums	See list of Scout badges in Chapter 4
Boys & Girls Clubs	After-school programs for children and teens
Clinics and Hospitals	Tours, labs, expert speakers
Boy Scouts	See list of Scout badges in Chapter 4
Explore your Future www.CanTeenGirl.org	Tour Your Future, healthcare facilities
Foundations	City, Educational, Regional, National
Girl Scouts	See list of Scout badges in Chapter 4
Girls Who Code https://girlswhocode.com/	Introduces girls to coding and promotes interest in computer science
Introduce a Girl to Engineering Day events: https://www.wpi.edu/news/calendar/events/introduce-girl-engineering-day	Engineering program similar to Take Your Daughter to Work day. Check with local universities who have engineering programs.
Museums	Art, Computing, Engineering, Natural History, Science (See Appendix B listing)
Paint Companies	Chemistry and Manufacturing
SAM Academy Wheels	Fresno, CA Science, music, art
SECME, Inc.	Strategic alliance partner schools, universities, govt agencies and industry prepare minority youth for STEM careers
Society of Women Engineers: http://societyofwomenengineers.swe.org/	Many programs and scholarships for girls.
Technology Companies	Software developers, Computer programming, Computer repair
YMCA	Programs vary
YWCA	Programs vary
Zoos	Biology, Zoology

Figure 3.1: Partner Agencies

(Flowers, 1998, 115). Find the supervisor whose job is external communications and enlist his or her help. There may be procedures in place that require that all communications flow through this person. Be sure that you are in contact with the communications person as well as your manager/principal before inviting outsiders into your library.

Once the project or event is done, write up your story and submit it to the appropriate managers before sending it to multiple venues. Administrators need to know what will be said about the library and they may be able to direct your story to a personal contact. Don't forget that library publications are always looking for stories to share. Whether your story is a success story or a story of the pitfalls you encountered both will benefit the library community. Prepare a presentation for local, regional, state, provincial, or national library conferences. Librarians excel at sharing information, techniques, and projects so share your knowledge and your difficulties as well as you successes (Pandora, 2013, 171–172). You might be someone else's inspiration.

OTHER LIBRARIES
Private Schools

Private schools, whether independent, charter, or parochial are another avenue of collaboration for both public schools and libraries. Since public libraries serve multiple schools, homeschool families, and multiple communities, teen librarians may already have contacts within each entity and may know the librarians by name. High schools often have career counselors whose specialty is to locate career shadowing experiences or internships for their students. These specialists, and many private schools, maintain a database of contacts who can provide advice and support or who would be willing to serve as mentors. Collaborations, however, aren't always about money; they are often about sharing. Jointly sponsored juried art shows, math or chess competitions, even gaming contests, or code-a-thons can be sources of collaboration (Squires, 2009, 44). What better place to showcase such an exhibition than the school or public library?

University Libraries

After cementing a relationship with your counterpart at the local public or school library, turn your attention to other libraries near you including special or academic libraries (Cooksey, 2017, 72). Local universities often have First Year Experience (FYE) Librarians whose job is to aid first-year students of any age with their adjustment to college-level resources and expectations. Inquire as to whether your students have borrowing privileges (Flowers, 1998, 96). Advanced placement and post-secondary option college classes (sometimes paid for by state legislatures) provide students with the ability to take college classes while still in high school and earn college credits for them. Many FYE librarians welcome school groups to their universities as a means of recruiting future students and preparing all students for college work. Usually the FYE librarian cannot come to schools in person, but Skype, videoconferencing, and FaceTime can still add value as a substitute for face-to-face contact.

In Our Experience

Previous students who were taking these post-secondary classes often returned for assistance as they were too embarrassed to ask for help in navigating unfamiliar databases at the university.

Students who are off-campus to attend technical schools (also known as career centers or vocational schools) are often bussed to the facility with little time to explore resources there. The courses for highly technical fields offered there require specialty resources in the areas of information technology or auto or boat mechanics and their librarians can help you to meet the needs of the students you share. You may garner information about the issues students are having with their vocational assignments at the home school, so the conversation can go both ways. The vocational librarian can address technical needs, and the school and pubic librarians can help translate and convey problems and successes students are having with projects.

Special Libraries

Local special libraries include many STEAM fields including manufacturing, business, scientific, governmental, medical libraries, and museums. Their librarians have specialized skills, and often advanced degrees in their own fields and are potential speakers. For example, a company producing paint could supply tours as well as chemists who could assist on a program; an accounting firm could advise your Math Counts Club before a competition and an investment broker or firm could explain stock market or economic trends to your high school business students. Government agencies, such as NASA or the EPA have technical expertise that can be tapped for that special patron who wants to go beyond the curriculum and holdings of the home libraries. Museums, whether art, science, natural history, or computing often have libraries of their own and a speaker's bureau to help you to expand the knowledge of your teens. Refer to Appendixes A and B for listings of art, science, and technology museums and planetariums. These experts may already serve local libraries by serving on advisory committees, such as technology committees.

A reliable source for special, governmental, and medical libraries in the United States and Canada is the *American Library Directory*. You can search the print copy by state or province and then by city. Libraries are coded by type, for example, S=Special (includes corporate libraries), M=Medical, and G=Governmental. Each entry includes library expenditures, subject interests, special services, and branches. If print copies aren't available then sign up for a 14-day free trial that allows you to retrieve partial records (http://www.americanlibrarydirectory.com/). Utilizing search terms such as science libraries or business libraries will provide you with a library's name, address, phone numbers, website, e-mail, and special collection information.

Library Colleagues

When beginning a new project, your first thought is to determine the library experts in the field who can help you. You will likely feel like a detective as you search for those experts in the literature, through your own professional groups and learning networks, and through various conversations. You will also likely discover that there are a plethora of local experts in your city, region, and state, many of whom you can follow through social media and other electronic means. If you follow the LM_Net, PUB_YAC, or YALSA listservs or other discussion lists, you will have expanded your reach to national and even international librarians who have a unique perspective on the issues that you share.

VOLUNTEERS

The Need for Advisors

Discuss with staff and young adult patrons the popular local clubs in your area. Do students belong to the local 4-H Club, scouting organizations, or a Girls Who Code club? Do your middle school students head to the Y or Boys and Girls Clubs for after-school programs and supervision? (Refer to Figure 3.1 for a listing of additional clubs and groups.) Do they come to the public library to "hang out" until a parent can pick them up? Any of these scenarios provide you with an audience for an after-school STEAM program as well as a need for advisors and volunteers to help set up and run the programs. Reach out to club advisors and the staff at local after-school programs. They may be happy to collaborate with you on programs. Figure 3.2 provides a list of professional associations that may be able to provide advice.

General
AAUW American Association of University Women: http://www.aauw.org/
STEM Georgia: www.stemgeorgia.org
STEM Education Coalition, Washington, D. C.: http://www.stemedcoalition.org/
STEM Georgia: http://stemgeorgia.org/
U. S. Patent and Trademark Office: http://www.uspto.gov

Arts
Association of Art Museum Directors (AAMD): https://aamd.org/
Art Film Institute: http://www.afi.com/
Sundance Institute: http://www.sundance.org/
Recording Industry Association of America: http://www.riaa.com
Women in Film: https://womeninfilm.org/

Computers and Technology
Computer and Information Technology Occupations: http://www.bls.gov/
ooh/computer-and-information-technology/home.htm
Cyber Careers—National Security Agency: https://www.intelligencecareers
.gov/NSA/nsacyber.html
Entertainment Software Assocation: http://www.theesa.com
Institute of Electrical and Electronics Engineering Computer Society (IEEE):
http://www.ieee.org (Has a Women in Engineering component)
International Game Developers Association: http://www.igda.org
National Center for Women & Information Technology: https://www.ncwit.org/

Engineering
American Society for Engineering Education: http://www.asee.org and
http://www.engineeringk12.org
Institute of Electrical and Electronics Engineers (IEEE): http://www.ieee.org
National Action Council for Minorities in Engineering (NACME): http://www
.nacme.org/
Society for Hispanic Professional Engineers: http://shpe.org/
Society of Women Engineers: http://societyofwomenengineers.swe.org/
Women in Science and Engineering (WISE) (African American women):
http://www.birmingham.ac.uk/university/colleges/eps/news/student/
WISE-March-2015.aspx

Math
Mathematical Science Research Institute: http://www.msri.org/web/cms
National Council of Teachers of Mathematics (NCTM): http://www.nctm.org/

Science
Alpha Chi Sigma (AXE) Chemistry: https://www.alphachisigma.org/
(Fraternity for Men and Women in Chemistry)
American Indian Science and Engineering Society (AISES): http://www.aises.org/
Association for Women in Science: http://www.awis.org/
National Science Teachers Association (NSTA): http://www.nsta.org/
National Society of Black Physicians (NSBP): http://nsbp.org/
Society for American Chicanos and Native American Scientists: http://www
.msri.org/web/cms
Women in Physics (American Physical Society): http://www.aps.org/programs/
women/index.cfm

Figure 3.2: Professional Associations

Parents

Start close to home when looking for experts. Ask to speak at the school's PTA meeting. At the meeting, ask parents to help you to compile a listing of parents in STEAM fields explaining that you are looking for mentors to assist you in providing the best experiences for their children and that such assistance will require a very small time commitment. If your school maintains a database of parents, perhaps a secretary can lead you to the names of those in STEAM fields. You will probably need permission from a manager to gain access to this information. If information is not granted, suggest that a line be added to the PTA newsletter and to the annual form (electronic or paper) asking parents if they would be willing to be a speaker; if the answer is yes ask for name, job title, and the contact e-mail or phone number they prefer. Such a request can also be made to members of the public library's Friends of the Library group through their newsletter.

Parents have always helped out in schools, as volunteers, as members of an advisory board, perhaps they would help with a program during National Library Week or Teen Tech Week. In every community there are those who are the "power parents" they have the ear of the principal, the manager, the superintendent, or the director. They may well be a good source of advice by sitting on one of your advisory boards (Flowers, 1998, 95). Your school, district, or library may also have a file of local contacts for speakers or programs that can be a resource.

Businesses and Manufacturing

Where will you find the advisors to help you to plan these programs? It's a good idea to tap local business leaders for this role. What are the most important industries in your region? Is there a particular kind of farming that dominates? What manufacturers and plants are nearby—steel, auto, tool and die, paint? Any of them can provide speakers and advisors for you and role models for your teen patrons. Determine which of your managers attends the local business group meetings, Chamber of Commerce, Rotary, Kiwanis, and other organizations. Chances are their rosters will list CEOs of local companies and executives from nonprofits and social service agencies. Search for leaders in STEAM fields who can advise you on setting up programs, tours, or who can supply you with speakers. Do not neglect to contact the manufacturers you discovered earlier. Most of their jobs require technical skills in math, science, engineering, and advanced training. Students need to see a wide range of STEAM careers as possibilities and to know that many of these jobs also pay well. The list of professional associations in Figure 3.2 will give you other ideas for places to look for advisors and mentors.

Community Members

Get to know your community and parents through participation in the Friends of the Library group, a parental advisory board, or your manager's or principal's advisory board. Look also for ways that community members and parents can participate (Flowers, 1995 95–96). Perhaps you need volunteers for the Read-a-thon in March honoring Dr. Seuss or an event to celebrate Poetry Month or National Library Week in April. Those who cannot volunteer on a regular basis may be willing to volunteer to help for a one-shot program. While PTA/PTOs and Friends of the Library are often solicited for monetary help, don't let your only contact with them be begging for help with your budgetary needs.

Providing services to your patrons often includes moving outside of your library. Whether you partner with the teen librarian at the public library, school librarians, or those at local colleges and businesses the time and effort it takes will be worthwhile. Besides gaining another point of view for your projects you have gathered another group of experts who can provide advice and serve as mentors to your staff and your patrons. Your work with outside groups also allows you to demonstrate your value to the community and provides you with the chance to gather more supporters for future levy campaigns. You will have a new network of like-minded people who can provide great benefits to your program and ultimately to your students.

REFERENCES

American Library Directory, 2013–2014. Medford, NJ: Information Today, 2013.

Braun, Linda W. "Everything Is Messy." *American Libraries* 46, no. 11/12 (November/December 2015): 58.

Cooksey, Ashley J. "Partnerships beyond Four Walls." *American Libraries* 48, no. 1/2 (January/February 2017): 72.

Flowers, Helen F. *Public Relations for School Library Media Programs; 500 Ways to Influence People and Win Friends for Your School Library Media Center.* New York: ALA Neal-Schuman Publishers, 1998.

Johnson, Abby. "Reach Out through Outreach." *American Libraries* 45, no. 11/12 (November, December 2011): 48.

Pandora, Cherie and Stacey Hayman. *Better Serving Teens through School Library-Public Library Collaboration.* Santa Barbara, CA: ABC-CLIO, 2013.Squires, Tasha. *Library Partnerships: Making Connections between School and Public Libraries.* Medford, NJ: Information Today, 2009.

Todaro, Julie Beth. "Community Collaborations at Work and in Practice Today: An A to Z Overview." In William Miller and Rita M. Pellen, eds. *Libraries beyond Their Institutions: Partnerships That Work.* Binghamton, NY: Haworth Information Press, 2005–2006. 137–156.

Wallace, Danny P. *Library Evaluation: A Casebook and Can-Do Guide.* Englewood, CO: Libraries Unlimited, 2001.

FURTHER READING

Doll, Carol A. *Collaboration and the School Library Media Specialist.* Lanham, MD: The Scarecrow Press, 2005.

Frye, Julie Marie, and Vaughn W. Nuest. "'The Mountains Are Calling' & You Must Go: Spending Part of Your Summer at a National Park Service Site." *Knowledge Quest* 43, no. 5 (May/June 2015): 56–58.

Miller, R.T. "We Need Tag-Team Leadership." *School Library Journal* 58, no. 5 (May 2012): 11. http://search.ebscohost.com/login.aspx?direct=true&db=tfh&AN=74998347&site=eds-live

Miller, R.T., and L. Girmscheid. "It Takes Two." *School Library Journal* 58, no. 5 (May 2012): 25–29. http://search.ebscohost.com/login.aspx?direct=true&db=f5h&AN=74998359&site=eds-live

Miller, William, and Rita M. Pellen, eds. *Libraries beyond Their Institutions: Partnerships That Work.* Binghamton, NY: Haworth Information Press, 2005–2006.

Murvosh, M. "Partners in Success." *School Library Journal* 59, no. 1 (January 2013): 22–28. http://search.ebscohost.com/login.aspx?direct=true&db=tfh&AN=84641701&site=eds-live

Smallwood, Carol, and Kim Becnel, eds. *Library Services for Multicultural Patrons: Strategies to Encourage Library Use.* Lanham, MD: Scarecrow Press, 2013.

4

◇ ◇ ◇

ENCOURAGING GIRLS
AND MINORITIES

ENCOURAGING GIRLS AND MINORITIES
TO PARTICIPATE IN THE SCIENCES

"According to the Census Bureau's 2009 American Community Survey (ACS), women comprise 48 percent of the U.S. workforce but just 24 percent of STEM workers" (Women in STEM, 2011, 2). Women with STEM jobs earned 33 percent more than comparable women in non-STEM jobs. As a result, the gender wage gap is smaller in STEM jobs than in non-STEM jobs (Women in STEM, 2011, 4). The survey explains that women fight stereotypes, have few role models, and find that the male-dominated fields are not flexible when it comes to family issues. As a bonus, women with college degrees who are working in STEM fields earn considerably more than women in other fields, as much as 20 percent more (Women in STEM, 2011, 7). In 2012 the Department of Labor reported that minorities are underrepresented in STEM fields accounting for only about 5 percent of STEM workers (Williams, 2014, 2).

How do we solve these shortages? What hurdles must women and minorities overcome? Once they are in STEAM fields, how do we encourage them to stay? First to define our term, we define minorities as those who are not Caucasian or Asian males since these two groups are the most dominant in the scientific fields, particularly, in graduate, engineering, and medical schools (Williams, 2014, 2).

WHY ENCOURAGE MINORITIES AND WOMEN?

First, let's deal with the question of why we should do so. For starters our country needs more qualified people in the fields of science, technology, computers, mathematics, engineering, and yes, art. Data from the U.S. Bureau of Labor Statistics (BLS) estimates that between 2012 and 2022, we'll need about a million more STEM workers (Vilorio, Dennis. 2014, 3). Economists now project that the United States will be graduating far fewer engineers and scientists than countries in the Far East (Williams, 2014, 2). In addition, the population of the United States is changing rapidly. About 91.7 percent of the population growth from 2000 to 2010 was from racial or ethnic minorities (PEW Research Center, *2011 Study: Minorities*). The 2010 U.S. Census reported that Latinos were the largest minority group living in the United States comprising 16.3 percent of the total population while African Americans counted for 12.2 percent of the population (PEW Research Center, *2011 Study: Minorities*). Furthermore, Williams states that "Blacks and Latino students make up less than 20% of those studying science or mathematic disciplines" (Williams, 2014, 2).

The Census Bureau also stated that STEM fields were overwhelmingly staffed by males (U.S. Census Bureau, 2014, 1). Additionally, a 2013 American Community Survey report noted that many women quit their STEM jobs after 10 years. So not only do we have job opportunities but we also have large numbers of women leaving the field early in their careers. Our questions expand to once we have women and minorities in STEM fields, what do we do to keep them there?

OBSTACLES

Vocabulary

One of the barriers that we need to remove is embedded in the terminology that we use to describe STEM courses. If we refer to courses as "rigorous or advanced" we scare students away, especially girls and minorities (Belser, 2015, 2). If these students also come from a socially disadvantaged area (i.e., less affluent school districts), or if parents aren't supportive, they may fear that they don't have the emotional or financial support to continue in these courses or these careers. Since many STEM careers require advanced degrees, for example, medicine, engineering, coding, programming, or research, the support of parents, guardians, counselors, and teachers is crucial to success in STEM coursework.

Instead, we need to emphasize the advantages of taking these courses and the opportunities that are open to students who have a good background in STEM classes. No longer is it sufficient to take only two years of high school math and science as was required in the 1960s and 1970s. Indeed, many states now require far more coursework in these areas and have moved the Algebra I course (in past years the first high school math class) earlier to the level of 7th or 8th grade.

If instead of labeling such coursework in terms of difficulty we emphasize the collaboration and problem-solving aspects of these careers and we describe the courses as in terms of creativity and teamwork then we make the classes much more attractive to all students.

Desirable, Doable, and Intriguing

We need to put a positive spin on these courses and make them sound desirable, doable, and intriguing.

Adult Perceptions

The expectations of parents, teachers, and other adults make a great impact on teens and pre-teens. Girls and minorities must be encouraged to consider careers in the STEAM fields. One of the findings of the Carnegie Science Center survey was that "parents are intimidated by science and mathematics. They feel that if students aren't planning to go to college, they do not need to take advanced science or math classes in high school" (Belser, 2015, 2). At a time when girls and minority students are searching for a career path it's also important to encourage the adults in their lives to present a variety of pathways that include STEM/STEAM careers. You can start by searching for community partners to help us overcome these adult perceptions. Refer to Figure 3.1 Partner Agencies for groups that can help you to plan your programs.

Geography

Perhaps the mentors, museum curators, or speakers you've identified as possible collaborators are on the other side of the country or the other side of the world. Videoconferencing can help to solve the problem without a large outlay of cash. Skype and FaceTime are programs that students use to chat with friends and relatives. Just as author chats and interviews can often be done with these programs, reaching female scientists, engineers, artists, musicians, software developers, computer programmers, academics, and medical professionals can be arranged, often without cost.

If you wish to include a nonprofit agency or university in your discussion, it is worth checking to see if they have a web-based videoconferencing tool that you could use without cost. Note: While schools often had videoconferencing tools in the past these were often bulky pieces of equipment or were housed in an area far from the library and scheduled for online classes. Other learning tools that are showing promise for library use are Facebook via live or video chat (Gilliss, 2014, 48) and Google + Hangouts. Used often for meetings, professional development, and online book clubs this tool has the advantage of "going live" with multiple participants. This would allow your library to collaborate with public libraries, school libraries, cultural centers, and other agencies as you interact with your speaker. Young adults would then have the chance to ask questions directly of people in their field of interest. In addition to the online discussion these sessions can be recorded and archived for those who could not attend or for those who wish to review the session again at a later date (Gilliss, 2014, 47).

SOLUTIONS
Start Early

Here are ways we can combat these issues. We need to start including, and encouraging, girls and minority students to participate in the STEM arena as early as possible; elementary school is not too early (Editorial, 2016). We need to create "interest and awareness early enough to take the needed courses and solve the science education gap" (Daniel, 2014, 1). Let's create the awareness that the fields of math, engineering, technology, and the sciences are open to all genders and ethnic groups.

Family Math

Teachers in the elementary schools try to promote math studies as a pleasurable experience. A kindergarten teacher has students count off their steps by tens while moving through the halls to art class. They sponsor Family Math nights where families play math games and receive ideas for use at home. An art teacher has students recreate fractal patterns they see during their nature walks around the school. Parents give their children the task of counting subway stops until they reach their destination. When with their friends they may learn how to calculate the unit cost of slices of pizza.

In Our Experience

Inspired by the family math concept I devised the "cash register game." If my sons could calculate the cash I would receive before the cash register noted it, they would receive the coins (not the dollar bills). They loved the game and became quite proficient at doing math in their heads.

Role Models

One of the problems is that young adults rarely see professionals from STEAM fields in the course of their school day. Unless their parents or some other adult they know happens to be a scientist or engineer or other STEAM professional, they are missing role models who help them to visualize themselves in the role of a pharmacist, a jet pilot, an electrical engineer, an accountant, or a computer programmer. You can remedy this by bringing in speakers who look like your young patrons from local companies, professional organizations, and medical facilities. See Figure 4.1 for role models in these fields; see also Figure 3.2 for a listing of groups that may be able to provide speakers.

STEM and STEAM Women		STEAM Field
Barton, Clara	Nurse in Civil War	Medicine
Blackwell, Elizabeth	1st American woman to study medicine at university level	Medicine
Cassatt, Mary	Impressionist	Artist
Cooney, Joan Ganz	Children's Television Workshop; *Sesame Street*	Art and Video Production
Copland, Misty	1st African American prima ballerina	Art
Curie, Marie	Two Nobel prizes	Chemistry and Physics

Figure 4.1: STEM and STEAM Women

Earhart, Amelia	1st female to cross Atlantic; Aviatrix	Science and Math
Easley, Annie J.	African American computer scientist, mathematician	Computer Science, Science, Math, Rocket Science
Fossey, Dian	Studied gorillas	Science
Franklin, Rosalind	Studied DNA and RNA	Chemist
Goodall, Jane	Studied chimpanzees, primatologist, Tanzania in 1960	Science
Herschel, Caroline	German astronomer who discovered comets, early 1800s	Astronomy
Hopper, Admiral Grace	Computer programmer, 1944	Technology
Jamieson, Mae	African American astronaut	Science
Jamison, Judith	Dancer	Artist
Kahlo, Frida	Painter	Artist
Kwolek, Stephanie	Invented Kevlar	Science
Lamarr, Hedy	Communications; led to smartphones	Science
Lange, Dorothea	Photographer	Art
Lin, Maya	Sculptor, Vietnam Wall	Art
Lovelace, Ada	1st female computer programmer, 1800s; created the first algorithm	Mathematics and Technology
Mitchell, Maria	1st professional astronomer; 2nd woman to discover comet; 1st woman, American Academy Arts & Science in 1848	Astronomer
Moses, Grandma	Painter	Art
Nightingale, Florence	Nurse in Crimean War	Medicine
O'Keefe, Georgia	Painter	Art
Reichs, Kathy	Forensic anthropologist, United States and Canada	Medical
Ride, Sally	1st American female astronaut; June 18, 1983, PhD, Physics	Science
Windsor, Edith "Edie"	1st female computer programmer at IBM; senior systems programmer	Computer science

Figure 4.1 (*continued*)

From *Full STEAM Ahead: Science, Technology, Engineering, Art, and Mathematics in Library Programs and Collections* by Cherie P. Pandora and Kathy Fredrick. Santa Barbara, CA: Libraries Unlimited. Copyright © 2017.

Introduce teens to literature that details the careers and lives of women and minority STEAM workers. The book, *Hidden Figures* (and the movie by the same name) relates the stories of talented African American women who worked for NASA in the 1960s serving as human computers. Their stories, and they, were hidden from the history books even though their calculations put men into orbit and on the moon. In her book author Margot Lee Shetterly demonstrates the power of role models to influence our interests, the courses that we take, and the careers that we pursue (Shetterly, 2016, p. xiii).

> "As a child, however, I knew so many African Americans working in science, math, and engineering that I thought that's just what black folks did."
> (Margot Lee Shetterly, *Hidden Figures*, Used with permission)

Take them online to visit sites such as the "The Immutable Impact of Black Scientists and Inventors" (https://www.edutopia.org/blog/impact-of-black-scientists-inventors-ainissa-ramirez) by Ainissa Ramirez. There are also societies and professional organizations for women and minority groups in many of the STEAM areas such as the Society for Hispanic Professional Engineers (http://shpe.org/). See Figure 4.2 for a listing of STEM and STEAM men role models; see further examples in the Resources section at the end of this chapter.

STEM and STEAM Men		STEAM Field
Banneker, Benjamin	Surveyor; helped to plan the District of Columbia	Science
Bell, Alexander Graham	Inventor	Communications
Brush, Charles F.	Inventor	Engineer
Carson, Ben	African American surgeon	Medicine
Carver, George Washington	Inventor and farmer	Botany
Cousteau, Jacques	Explorer; underwater explorer; conservationist	Science; Biology; Zoology
Crick, Francis	Nobel Prize for work in genetics	Biology
Dali, Salvador	Painter	Artist

Figure 4.2: STEM and STEAM Men

Davis Jr., Benjamin O.	African American general and commander of the Tuskegee Airmen during World War II	Aviation
Edison, Thomas	Inventor	Science and Physics
Einstein, Albert	Theory of relativity	Science and Physics
Fleming, Alexander	Discovered penicillin	Medicine
Friedman, Milton	Nobel Prize in Economic Sciences	Economist
Goddard, Robert	Rocket scientist	Astronomy and Engineering
Monet, Claude	Impressionist painter	Artist
Morgan, Garrett	Inventor; early gas mask	Engineering
Nash, John	Game theory	Economist
Oppenheimer, Robert	Rocket scientist	Astronomy and Engineering
Parks, Gordon	African American photographer during the Depression	Artist
Pei, I.M.	Architect of the Louvre and the Rock and Roll Hall of Fame	Architecture, Art
Picasso, Pablo	Painter	Artist
Renoir, Pierre-Auguste	Painter	Artist
Rivera, Diego	Painter	Art
Sabin, Albert	Developed oral polio vaccine	Medicine
Salk, Jonas	Developed polio vaccine (shot)	Medicine
Smith, Adam	*Wealth of Nations*	Economist
Tesla, Nicolai	Inventor	Science and Physics
van Rijn, Rembrandt Harmenszoon	Painter	Artist
Von Braun, Wernher	Former Nazi rocket scientist granted asylum	Astronomy and Engineering
Watson, James	Nobel Prize for work in genetics	Biology

Figure 4.2 (*continued*)

Online resources provide other opportunities including A Mighty Girl and the National Girls Collaborative Project (see websites in the Resources section at the end of this chapter). Another resource for finding experts willing to assist you is the FabFEMs: Find a Role Model web page (http://www.fabfems.org/). Women, including those of color, list the activities that they will provide including after-school programs, hands-on activities, and field trips. On their page you can search by STEAM specialty, for example, Aeronautics, Information Technology, or Materials Research and filter by city or state (using postal code abbreviations). Don't limit yourself to those who are local or within driving distance. Due to social and electronic media apps you have many ways to connect without being in the same building.

While it would be best to have STEAM professionals in-person as role models, it isn't always possible. When you can't bring in a face-to-face speaker, another option is to use an app like Skype or FaceTime or videoconferencing. Promote women and minorities through your displays and collections. Role models need not be physically in your library to be of service to your pre-teens and high school students. During her research on *Hidden Figures*, Shetterly compiled a great deal of information on these human computers and is turning them into a database to keep their stories alive. In conjunction with Macalester College, the alma mater of many of these women, Shetterly created the *Human Computers at NASA Project Digital Archives* (http://omeka.macalester.edu/humancomputerproject/about). Start here to compile information about African American women for your displays.

While you are planning displays consider having a showcase or bulletin board with a revolving display of STEAM professionals. Highlight scientists one month, mathematicians the next, and artists the next until you have displayed careers for each STEAM acronym. You need not wait for the typical Black History (February) or Women's History (March) Months to celebrate their unique contributions. However, if your library traditionally does so, here are a few additional months to celebrate:

Asian-Pacific American Heritage Month (May): http://www.apples 4theteacher.com/holidays/asian-pacific-american-heritage/when-is-asian-pacific-heritage-month.html

National Hispanic Heritage Month (September): http://hispaniche ritagemonth.gov/

Native American Heritage Month (November): http://nativeamer icanheritagemonth.gov/

After-School Programs

Physically bringing in speakers from local companies, universities, and agencies is a great way to expose girls and minorities to the STEM fields. It is especially helpful if you can provide diverse speakers who look like your audience. Williams reports that the Albany College of Pharmaceutical and Health Sciences (ACPHS) created a program for minority students in grades 3 through 8. Problem-solving exercises and demonstrations are held on their campus and have included robotics and experiments with water balloons (Williams, 2014, 2). In Miami, Florida, in 2005 a program called Experience Aviation was geared to African American students. Its goal was to allow them to see the possibilities in the sciences and, particularly, in the field of aviation (Williams, 2014, 3).

In addition to speakers, many museums, zoos, and science centers have mobile labs that visit schools. Fresno has a SAM Academy on Wheels (Science, Art, and Music) which travels to reach migrant Latino workers. Columbus, Ohio has a Center of Science and Industry (COSI) museum mobile lab and the MdBio Foundation has a mobile science lab that travels to schools in Maryland (Daniel, 2014, 2). Even sports such as NASCAR hold seminars to educate students in the "practical side of science and math."

Summer Programs

An organization called the Level Playing Field Institute, or LPFI, created the SMASH Academy to assist pre-teens and teenagers in Oakland and San Francisco. SMASH Academy enlists the help of faculty members from four nearby universities to help students to experience technology, the sciences, and social issues by creating apps that help others, for example, job search or calorie counting. Professors from Stanford University, the University of California Los Angeles, the University of Southern California, and the University of California—Berkeley helped with the summer program. When few African American males applied they created another preparatory program called SMASH Prep that meets twice a month to encourage enrollment (Williams, 2014, 3–4).

Summer programs are often run by scouting organizations, the YMCA and YWCA, and the Boys and Girls Clubs of America. Investigate those agencies that exist in your areas. Many of these organizations sponsor programs that fall into the STEAM fields. For program ideas, check the list of STEAM Scout Badges in Figure 4.3.

STEAM Field	Boy Scout Badges	Girl Scout Badges—Juniors	Girl Scout Badges—Cadettes	Girl Scout Badges—Seniors

Science	Animal Science
	Astronomy
	Automotive
	  Maintenance
	Aviation
	Bird Study
	Chemistry
	Dentistry
	Electricity
	Electronics
	Energy
	Environmental
	  Science
	Fingerprinting
	First Aid
	Fish & Wildlife
	  Management
	Forestry
	Geology
	Insect Study
	Mammal Study
	Medicine
	Metalwork
	Mining in Society
	Nature
	Nuclear Science
	Oceanography
	Plant Science
	Public Health
	Reptile &
	  Amphibian Study
	Robotics
	Soil & Water
	  Conservation
	Space Exploration
	Veterinary
	  Medicine
	Weather
	Welding
		Health Aid	Animal
			Kingdom
			(Birds)
			
			Animal
			Kingdom
			(Insects)
			
			Naturalist—
			Trees
			
			Star
				Naturalist—Sky
				
				Science of Style
				
				Women's
				Health
				
				Health and
				Fitness Projects
				
				Natural
				Surroundings/
				Camping
				Projects

Technology	Communication Digital Technology Game Design Movie Making Programming		Digital Movie Maker	Website Designer

Figure 4.3: STEAM Scout Badges

From *Full STEAM Ahead: Science, Technology, Engineering, Art, and Mathematics in Library Programs and Collections* by Cherie P. Pandora and Kathy Fredrick. Santa Barbara, CA: Libraries Unlimited. Copyright © 2017.

Engineering	Architecture Engineering Inventing Model Design & Building Robotics		Metal Arts	
Art	Animation Art Basketry Bugling Drafting Game Design Graphic Arts Music Painting Photography Pottery Sculpture Textile Theater Wood Carving	Art in the Round Dancer Drawing & Painting Musician Prints Songster Weaving & Basketry	Photography	Troupe Performer Arts & Humanities Projects
Mathematics	American Business Entrepreneurship Personal Management		Entrepreneur	Math, Science, & Technology Projects

Figure 4.3 (*continued*)

STEM Camps

Local colleges often hold their own STEM camps, for example, Blooms-burg University in Pennsylvania holds weeklong day camps for students called Great STEM Adventure Camps. The camps are broken into three divisions by ages they serve, including students in grades 5 through 10. Bloomsburg started their camps as they saw the need to encourage younger students to view STEM careers as an option for study in col-lege. Their research told them that Pennsylvania primary school students receive less than three hours of instruction weekly in the sciences while 78 percent of 8th graders receive less than five hours of such instruction (http://www.bloomu.edu/stem-programming).

In 2015 Bloomsburg held their first GI-STEM, Girls in Science, Technol-ogy, Engineering, and Math day at their Regional STEM Education Center. One hundred Girl Scouts representing 21 different troops worked through 11 different learning stations geared to STEM topics. Numerous commu-nities hold Imagine STEM summer camps that introduce Girl Scouts to local female scientists. They are always looking for volunteers to help—you need not be working in a STEM field, just have an interest in bringing

STEM awareness to girls. Check local websites for other STEM-oriented camps (e.g., http://www.girlscouts.org/en/our-program/ways-to-partici pate/series/imagine-stem.html).

Tours

One of the options that can help girls and minority students to view STEAM careers as a viable choice is to provide the opportunity to tour labs and manufacturing sites to meet experts in the fields.

Belser mentions a Pittsburgh area program called Tour Your Future designed for pre-teen and teenage girls between the ages of 11 and 17. Tours were done outside the school day and allowed girls to visit the labs at the University of Pittsburgh and Carnegie Mellon University as well as museums and local hospitals in 2015. She reported that Pittsburgh area companies like Alcoa and Valspar welcomed the young visitors to explain how science and math were involved in their manufacturing processes (Belser, 2015, 2).

Don't forget that manufacturing jobs require a great deal of scientific, mathematical, and technological skills, that workers are in short supply, and that experienced workers earn salaries commensurate with other STEM fields, for example, $69,454 for experienced workers (Youngstown, 2013, 1). In 2013 the YWCA of Youngstown created a one-week program for girls in 4th through 6th grades. Support from the United Way and a private foundation supplied the seed money for their endeavor. Besides visits to factories girls were presented with challenges that included designing something that would be presented at the end of the week (Youngstown, 2013)

LEGISLATION

On March 17, 2016, Senate bill 2710 the Women and Minorities in STEM Booster Act of 2016 was introduced by Senator Patty Murray of Washington State (Seattle Schools Community Forum). The intent was to provide money for National Science Foundation grants in the areas of STEM for "outreach, mentoring, and professional development" and to encourage girls and minorities to enter STEM fields and to retain them once they are in the field (Senate Bill 2710). Unfortunately, it died in committee and was not put into law. There were inroads made in 2016 as STEM was a priority. See the blog "STEM for all" (https://www.whitehouse.gov/blog/2016/02/11/stem-all). We hope that current and future leaders and legislators will take up the mantle of improving STEAM education.

PARTNERS

Some of the first groups to search for are social service agencies and local cultural centers in our communities, that is, those that provide services to the African American, Hispanic, Asian, or LGBTQ communities. Meet their leaders to learn more about the service they provide to pre-teens and teens. In searching for some common ground for future speakers, mentors, and programs you can add additional value for your patrons or promote services that do not yet exist (Stripling, 2014, 5). To transform your community toward greater acceptance of possibilities for all patrons try to serve as an ambassador or liaison between your library and these groups (Fiels, 2014, 6).

Gather demographics for your area. What are you minority populations? Are these long-term residents or new arrivals who may need help transitioning or overcoming language barriers? Are new businesses planned that will draw workers or managers from abroad? Your director, superintendent, or manager may be a member of the local Chamber of Commerce and privy to this information. Whatever the results, learn about the culture of these groups. If this is a well-established group you have probably already added resources to the library to assist them or have learned where you can borrow materials in their native language. The problem lies in the fact that many people of minority cultures gain information through informal means via those who are also native speakers or through religious organizations such as churches, temples, or mosques (Ruhlmann, 2014, 38). Some may be distrustful of governmental institutions based on their experiences in their home countries. Second, those who have emigrated from foreign countries do not expect the breadth of material that we can provide to them. They may have had little access to libraries due to distance or limited hours. They don't anticipate finding sources to help them to learn English, get their GED, apply to graduate school, study for citizenship, driver's licenses, or military exams, renew passports, or obtain tax forms. If individuals are undocumented, they generally have a healthy fear of institutions, including libraries (Ruhlmann, 2014, 38–39).

One of the first things that you can do is to look for help from outside groups, particularly those who also enjoy working with young adults. Let us look first at two organizations geared specifically to pre-teens and teens—the scouting organizations. If your library could provide space occasionally for or provide outreach to the scouting organizations in the cities that you serve you would reach an additional audience. For younger scouts often the parents take turns providing the evening's program borrowing resources from local sources and researching the subject to provide background information for the scouts. Figure 4.3 gives you a listing of STEAM Scout badges to help you plan programs. (Additional information about potential partners can be found in Figure 3.2.)

Notes from the Field

Check with your local museums for tangible objects that you can display or use in your programs. Our local natural history museum allows teachers to borrow materials from the Teacher Resource room, for example, rocks, for school use or scout meetings. The art museum brought out artifacts for school and library display cases.

Boys and Girls Clubs and other local after-school programs may benefit from your offer of outreach programming. Many create thematic programs for each month or quarter of their calendar; covering one of the STEM/STEAM topics may be a welcome addition to their schedule.

Also contact local adult organizations such as the Chamber of Commerce, Rotary, and Kiwanis as they often provide many hours of service to local schools and youth organizations. Chances are that there will be men and women in these organizations who are involved in STEAM careers who would be willing to be a speaker, consultant, or advisor to you. Try to find speakers who are female, members of minority groups, or both. Teens need to envision themselves in these careers. Other organizations that are active in your community should be researched as well. In my community the Garden Club and League of Women Voters were very active in providing scholarship monies to the schools and may have women who are willing to speak.

RESOURCES

A Mighty Girl. http://www.amightygirl.com/. Website, Facebook, and social media presence. Books and lists of "Mighty girls" and women. Excellent resource for working with young women.

Astro4Girls. http://www.ala.org/programming/astro4girls-resources. A collection of websites full of Astronomy sites and STEM resources.

Best Apps for Teaching and Learning. http://www.ala.org/aasl/standards/best/apps/2016

FabFEMs: Find a Role Model. http://www.fabfems.org/ Resource guide for female role models to speak to or connect with young people. Guide gives the area of expertise. Includes women of color.

Human Computers at NASA Project Digital Archives. http://omeka.macalester.edu/
    humancomputerproject/about
National Girls Collaborative Project. http://ngcproject.org/statistics Statistics,
    mini-grants, webinars and resources for girls and people helping girls
SECME, Inc. http://secme.secme.org/ Strategic alliance of partner schools, uni-
    versities, government agencies and industry. They push for diversity and
    prepare minority youth for STEM careers through summer programs.
STEM Education Coalition, Washington, DC. http://www.stemedcoalition.org/
STEM Georgia. http://stemgeorgia.org/. Has information about online educa-
    tional and career resources for their residents, competitions, schools, and
    underrepresented groups—such as women and minorities.
Society for Hispanic Professional Engineers. http://shpe.org/ Source for network-
    ing. At their 2016 conference they sponsored a Hackathon for students.
Society of Women Engineers, http://societyofwomenengineers.swe.org/. Offers
    program outreach to schools, and scholarships and awards.
Women in Science and Engineering (WISE). http://www.birmingham.ac.uk/uni
    versity/colleges/eps/news/student/WISE-March-2015.aspx. High achiev
    ing African American women pursue degrees in physical science. Mentor
    and support minority women with a joint engineering program at Spelman
    College and Georgia Tech.

REFERENCES

Belser, Ann. "Exposing Students to STEM Fields Early Helps Girls, Minorities
    See Potential." *Pittsburgh Post-Gazette (PA)*, November 17, 2015.
Daniel, J., Abdun-Nabi, and Finkelstein, J.J. "Minority Students Are the Future
    of STEM." *Washington Post,* January 3, 2014.
Editorial. "Effort to Spur Women, Minorities to Pursue STEM Careers Wor-
    thy." *Walla Walla Union-Bulletin (WA),* March 23, 2016. *Newspaper Source,*
    EBSCOhost.
Fiels, Keith Michael. "Libraries Transforming Communities." *American Libraries*
    45, no. 5 (May 2014): 6.
Gilliss, Apryl Flynn. "A Novel Idea." *American Libraries* 45, no. 5 (May 2014): 45–49.
PEW Research Center. "Minorities Account for Nearly All U.S. Population
    Growth." ACTANK: News in the Numbers. March 30, 2011, p. 1. http://
    www.pewresearch.org/fact-tank/2011/03/30/minorities-account-for-nearly-
    all-u-s-population-growth/
Ramirez, Ainissa. "The Immutable Impact of Black Scientists and Inventors." *Edu-
    topia: Technology Integration.* Blog. February 12, 2014. https://www.edutopia
    .org/blog/impact-of-black-scientists-inventors-ainissa-ramirez
Ruhlmann, Ellyn. "Connecting Latinos with Libraries." *American Libraries* 45, no.5
    (May 2014): 36–40.
Shetterly, Margot Lee. *Hidden Figures.* New York: William Morrow, HarperCollins,
    2016.
Stripling, Barbara K. "Equity, Diversity, and Inclusion." *American Libraries* 45,
    no. 5 (May 2014): 5.
U.S. Census Bureau. "Census Bureau Reports Majority of STEM College Gradu-
    ates Do Not Work in STEM Occupations." July 10, 2014.

Vilorio, Dennis. "STEM 101: Intro to Tomorrow's Jobs." *Occupational Outlook Handbook* (2014): 2–12. http://www.bls.gov/careeroutlook/2014/spring/art01.pdf

Williams, Joseph P. "Expanding the STEM Pipeline." *U.S. News Digital Weekly* 6, no. 23 (2014): 13.

"Women in STEM: A Gender Gap to Innovation." Washington, DC: U.S. Department of Commerce, Economics and Statistics Administration, August 3, 2011. http://www.esa.doc.gov/sites/default/files/womeninstemagaptoinnovation8311.pdf

"Youngstown STEM Workshop Targets Girls." *Vindicator* (Youngstown, OH), July 5, 2013. *Newspaper Source*, EBSCOhost.

FURTHER READING

California Department of Education. News Release. "State Schools Chief Tom Torlakson Marks Women and Girls in Science, Technology, Engineering, and Mathematics (STEM) Week." April 6, 2015.

Cunningham, Christine M. and Melissa Higgins. "Engineering FOR Everyone." *Educational Leadership* 72, no. 4 (2014): 42–47.

Exner, Rich. "Minorities Now 38.4% of Nation." *The Plain Dealer* (Cleveland, OH), June 24, 2016.

Slaughter, Louise. "Proof that the STEM Fields Are Totally Inhospitable to Women." *MS Magazine.* Blog. April 14, 2016. http://msmagazine.com/blog/2016/04/14/proof-that-the-stem-fields-are-totally-inhospitable-to-women/

Yerak, Becky. "STEM-inspired dolls in demand." *The Plain Dealer* (Cleveland, OH), December 20, 2015. P. F2.

5

◇ ◇ ◇

SCIENCE AND MATH

SPUTNIK

On October 4, 1957, the Soviet Union shocked the world when it launched the first artificial satellite, Sputnik I into orbit around the Earth. Nations felt that the Soviet had shown a spotlight on its superiority in math, science, and engineering. As a result the U.S. government infused large quantities of money into education specifically to improve the science and math skills of our youth. Today, as we approach the 60th anniversary of the Sputnik I launch we are in a similar situation. To meet the challenges we face, we need more workers qualified in the STEM fields. We need to provide rewarding experiences in science and math for our young people leading them to consider careers in those fields where employers are looking for STEM workers. This chapter details how we can do so within the curricular areas of math and science. You'll find information about activities that can be used in developing programs that meet the needs of your library. In addition, we have compiled lists of supplies, vendors, museums, and professional associations you can find in Appendixes A–D. Chapter 9 discusses how makerspaces are helping students to learn via hands-on activities, and learning to experiment without fearing failure or a poor grade.

LACK OF STEM WORKERS

Currently the United States lacks qualified STEM workers in many fields. The 2012 study undertaken by the President's Council of Advisors on Science and Technology declared that we needed a million more STEM workers within the next 10 years (Xue and Larson, Bureau of Labor Statistics, 2015). While jobs in academia do not seem to have trouble finding talented PhDs the research, manufacturing, and computing sectors lack talented jobseekers with doctoral degrees. Fields such as nuclear engineering and materials science suffer shortages as defense contractors are required to reserve those jobs for U.S. citizens due to the sensitive nature of the work (Xue and Larson, Bureau of Labor Statistics, 2015). In the computing fields there were often multiple jobs, as many as seven, for each graduate (Rothwell, 2012). In 2013 the demand for software developers was up 120 percent over 2012 (Lombardi, 2013). But research and software are not the only areas that lack talented employees. The manufacturing sector is also feeling the loss of qualified STEM workers since fewer young people train for skilled jobs as machinists and technicians. In 2011, manufacturers were surveyed and it was estimated that over 600,000 jobs were left without workers (Wright, 2013). We need young people to embrace design and experimentation as represented by the STEM fields. Currently only 12 percent of females hold patents or have received awards for their inventions while three times as many patents (35.5%) have been awarded to those who are foreign born (Nager et al., 2016).

Since science and math subject matter and activities are co-dependent and overlap with activities often requiring both science and mathematics components they are discussed together in this chapter to avoid being too repetitive. In both subject areas we must teach our students not only to think critically but also to become problem-solvers as these are the two skills listed most often as desirable by business executives (November 2016, 2). Traditionally these two subjects were taught in isolation; today teachers employ a more integrated approach utilizing both curricular areas. To do so we need help. At best, librarians—whether school or public library—may have had an undergraduate degree in math or science. Unless a librarian has taught in either discipline, neither type of librarian is in a science or math classroom on a daily basis. We need, therefore, to collaborate with classroom teachers, to brainstorm, and embrace "out of the box" thinking. This also provides us with a good opportunity to model teamwork and lifelong learning for young people! We need to help our patrons to learn these skills while having fun at the same time.

In this chapter, you'll learn how you can do so within the areas of math and science. Start providing more inquiry learning and hands-on activities.

To make these disciplines even more engaging, focus on developing critical thinking, along with a willingness to give young people a voice in what is done. As a professional, acquaint yourself with the obstacles, as well as possible solutions, culminating in a discussion of activities and resources that help us build experiences that lead to student involvement and achievement.

Critical Thinking Skills

Young people who don't feel strong in math or science skills are less likely to have the problem-solving skills and logical thinking patterns to pursue these fields in college, leading to shortages of qualified candidates. As librarians we are in a unique position to foster inquiry learning and research. Critical thinking skills are utilized in researching a topic. The process requires students to utilize the scientific method and reinforce what they have learned in the classroom. Students must develop a hypothesis—their plan of attack, aka their search strategy. They need to determine which search terms they will use and which databases, websites, indexes, and print materials will provide them with the best resources. Once they begin their search they must problem-solve and evaluate what they have gathered. Have they located materials that will answer their questions or solve their problems? What are they missing and where can it be found? They must continually reevaluate and adapt. Who is there to guide them in their quest? Their librarians at both public and school libraries can help them to refine their searches and lead them to government documents and advanced resources such as subject-specific databases.

Let Your Patrons Guide You

Take advantage of the latest interests that your readers, and your non-readers, show. Can you tap into the interest in dystopian literature with games (board, video, computer) or an activity wherein patrons create their own dystopian society and provide the scientific background for the choices they made? Does interest in sharks spike during TV's *Shark Week*? Are your teens lovers of campy shows such as *Sharknado*? Do your patrons watch Animal Planet, Nova, or Planet Earth? Capitalize on their interests. Perhaps you can have them design their own video show featuring their favorites, or local interest topics, such as recycling, animal shelters, flora and fauna, or bird watching. To learn more about students then can be seen through observation, try a survey, on paper or online, asking their interests. See Figure 5.1 as a starting point.

After listing your favorite activities, place check marks in the proper category.						
What I like to do	**Science**	**Technology**	**Engineering**	**Art**	**Math**	**Other? Please list the category**

Figure 5.1: Skills and Hobbies

From *Full STEAM Ahead: Science, Technology, Engineering, Art, and Mathematics in Library Programs and Collections* by Cherie P. Pandora and Kathy Fredrick. Santa Barbara, CA: Libraries Unlimited. Copyright © 2017.

If you don't already have an ongoing relationship with your local public or school librarians, then this is a great opportunity to take your first step. School librarians can alert teen and reference librarians to assigned curricular projects and papers, as well as summer reading titles. Teen and reference librarians can inform school librarians of new acquisitions, upcoming events, and author talks and share ideas raised in meetings. Both share interests raised from their Teen Advisory Boards. An added bonus is that by working with classroom teachers you will learn more about their specialties and their passions. Just by listening to them and what they're trying to accomplish you learn more about how best to connect with them and support them in their work. By communicating on a regular basis you can formulate ideas that dovetail with existing projects or work to create a new program. Compare notes on speakers that may be coming to either library and find ways to share the experience. This can also help with covering your costs.

Classroom Teachers as Patrons

As Common Core and state standards demand student literacy and reading in non-fiction works and scientific method, teachers also scramble to find appropriate resources and technology tools and need this type of help from librarians. In the past, lectures, solving equations, repeated practice, and experiments were the norm. Teachers are now expected to locate simulations, subject-specific databases and resources, and suitable non-fiction materials for their classes. While this often takes them out of their comfort zones, these are services that we can provide; they create the perfect opportunity to work with a new group of teachers. Classroom teachers also do not have the wealth of resources that we have to access reviews, nor do they have the time to search. For example, they may come to us with a title such as *The Immortal Life of Henrietta Lacks* by Rebecca Skloot (Broadway Books, 2011) and need to know if it is appropriate for their grade level.

Once you know more about their course of study and their passions you can alert them to materials that cross your desk. Plan a joint workshop for new teachers to alert them to resources at their new school and public libraries; perhaps you can demonstrate how to set alerts to search terms in their field of interest. Librarians continually work with the standards in our own profession such as AASL's *Standards for the 21st-Century Learner* or the Search Institute's "40 Developmental Assets for Adolescents (ages 12–18)." It also helps you become more aware of the standards that science and math teachers are dealing with on a daily basis. See relevant standards for math and science listed in the Further Reading section of this chapter and in Figure10.1. This is a challenging task but librarians always rise to the challenge; we create bibliographies and displays to support local teachers.

Obstacles to Student Engagement in Math and Science

How do you help students to overcome their fear of science and math? How do you overcome the rumors that these courses are hard and only suitable for the best and the brightest? How do you help students see the engaging side of math and science? While you might not think tweens and teens are old enough to say "remember when?" ask them to think back to the times when they enjoyed science and math. Those were the times when these subjects were fun and exciting, whether they were building architectural wonders with erector sets, counting candy pieces at Halloween, or ensuring that their sibling was cutting the birthday cake into equal pieces. Librarians, and teachers, often use portions of picture books as an introduction to new topics or programs in order to bring levity to the subject and to defuse the concern that students often show. Students react with smiles to favorite books like *The Math Curse* by Jon Scieszka and can relate to the example of waking *before* your alarm clock sounds only to count the minutes until the alarm goes off. Call this a library icebreaker; a way to get young people ready to jump into math and science activities.

One of the advantages of experiential learning is that patrons learn how to direct their own learning. They learn to work alone and to evaluate their own learning, says researcher John Hattie (November, 2016, p. 2). Students fear making mistakes, especially as they move into the middle school and high school years. They don't want to look foolish in front of their friends. Developmentally, the risk-taking child becomes a self-conscious young adult. We need to re-capture the spirit of the children who played video or computer games repeatedly until they found ways to reach the next level. These children didn't fear making mistakes, they learned from each failure and rejoiced at winning the next token, totem, or prize. We need to encourage risk taking. Math and science are good areas for this, as they utilize predictions, hypotheses, and experiments. We learn as much from failure as from success, and sometimes more.

> Failure is a necessary attribute of engineering. . . . That's quite a contrast with traditional schoolwork, where failure can carry a stigma. (Cunningham and Higgins, 2014, 46)

Thomas A. Edison is credited with saying that he hadn't failed, he had merely found 10,000 ways *not* to make a light bulb. While various sources change that number to 2,000 or 3,000 the premise is that we should always try, try, again (https://www.brainyquote.com/quotes/authors/t/ thomas_a_edison.html). Whether the subject is math, art, science, engineering, or computer technology we need to promote the idea that failure *is* an option, that the process is more important than the product. We learn from our mistakes. We experiment, analyze our errors, and try again. Practice might not lead to perfection but it will lead to stronger skills in problem-solving and analysis. And it will help students see that learning isn't always about that perfect grade, but about exploring and experimenting.

Solutions: Bringing Math and Science into the Library

Do you feel that your skill set doesn't include math and science? Contact others who may have an insight into activities, programs, and simulations that can help you. There are numerous STEM schools in both Canada and the United States. In addition there are many museums that plan activities for children of all ages. It may be a *Build It* day at the local science museum or a *Design Zone* exhibit at the Great Lakes Science Center (Cleveland, OH) that promotes hands-on activities like designing roller coasters, developing games, or producing music. Just as libraries connect with authors via videoconferencing why not Skype with a scientist, a mathematician, or an economist? Contact the Chamber of Commerce, for the names of local businesses that have Speakers Bureaus as a starting point. Governmental agencies, such as the Federal Reserve Bank or NASA, may provide speakers as well. Planning STEAM activities doesn't have to be complicated, try some of the ideas listed in Figure 5.2 and add your own ideas.

Investigate the science and math specialties in your region. Many cities are known for a specific specialty and are supported by local businesses and universities. New York City is a financial hub; Houston, Texas, is a space hub for astronaut training while Silicon Valley in California and Seattle (Washington) are technology hubs. For example, if you are in a rural area farmers and the local agricultural college can talk about the business of farming, the need for creating insect-resistant plants, and the effects of climate on the growing season. The local botanical garden can talk about plant hybridization.

For example, in Northeast Ohio two state universities located within a half hour of each other have each created their own niches. While they have varied curriculums that overlap, Kent State University specializes

Options	Examples
After-school programs	Experience Aviation (Florida), http://www.experience aviation.org/African-American students math/sci/ eng/aviation
Design Challenges	Robotics, Architecture Level Playing Field, solve problems, create apps
Girls Who Code Club	Hour of Code
Mentors	Minority & Female Engineers
Mobile labs	Zoo, Cosi, MdBioLab (Bioscience lab), SAM Academy on Wheels, http://www.fresnoartmuseum.org/ education/for-children/sam/
NASCAR educational seminars	Explore the "practical side of science & math"
Scholarships	NSA National Security Agency, e.g., Stokes Educational Scholarship Program, recruits high school minority students who major in computer science or electrical engineering
Summer camps with museums	See Appendix A: Art Museums and Appendix B: Science and Technology Museums and Planetariums
Summer camps with college partners	SMASH Academy for African-American & Latino Youth (Stanford, USC, UCLA, UC Berkeley)
Theory Classes	Music
Tours	Manufacturing companies, medical facilities
Workshops	Labs, museums, healthcare, art institutes

Figure 5.2: Encouraging STEAM

in liquid crystal display (LCD) technologies and the University of Akron specializes in polymer research. Nearby Cleveland has multiple private and public universities and businesses that support the industries of medicine, as well as biomedical, aerospace, and chemical engineering.

ACTIVITIES

Creating STEAM activities that interest teens can often be a stumbling block for librarians, many of whom graduated with degrees in the humanities and social sciences. Thankfully, there are numerous books and blogs available that provide programming ideas. Take time to enjoy searching the children's department for ideas as well. Books listing programs for younger audiences can be adapted for pre-teens and teens. Since librarians don't have to give grades, you can be freer in your choice of projects.

A children's book dealing with raptors can be followed by a presentation by an expert from the local zoo or natural history museum.

Many public libraries have homework centers that are structured with specific days and hours. They are staffed by tutors or librarians who assist students. Some libraries have joined the makerspace movement, which may offer opportunities for structured or unstructured learning. Think outside the box. What careers combine STEAM fields? Art, math, and science knowledge are required for careers in architecture, art conservation and preservation, and cartography to name a few that we don't typically consider to be STEAM career fields. Research the types of businesses and agencies in your immediate area. Discovering what is in your own backyard may lead you to the experts who can suggest activities. Since many companies promote volunteerism in their workers, you may be able to tap companies that don't have formal speakers bureaus. If many of your teen patrons are self-declared foodies, invite a local home economics teacher to speak about the chemistry of baking or have a chemist from a local paint company explain how new colors are produced and named. Bring in a chef to talk about the science of cooking. While older students have moved beyond the baking soda and vinegar volcanoes stage of their younger years they would probably enjoy taking part in similar experiments, especially if they don't have to follow it up with a lab report. Wouldn't it be wonderful to prepare a program for middle school students allowing them to design their own roller coasters?

Museums

While many libraries are located close to urban centers with museums that can be accessed via field trips or visits from curators and researchers not all libraries have this advantage. Museum websites often contain a wealth of information that can be useful for a program at your own building. Contact information allows you to get in touch with the specialist who could guide you in your mission. There may be a way to connect for free via videoconference calls, Skype, or FaceTime as you would with an author. Don't let distance be a hindrance to your vision.

See the listing of Art Museums and Science and Technology Museums and Planetariums in Appendices A and B. For additional resources visit Canadian writer Amy Leask's blog STEM, STEAM, and STREAM "20 American STEM Museums You Shouldn't Miss" (http://enable education.com/20-american-stem-museums-you-shouldnt-miss/) (Leask, May 6, 2013) and "Canadian STEM Museums, Coast to Coast" (http:// enableeducation.com/canadian-stem-museums-coast-to-coast/) (Leask, May 1, 2013).

Speaker Suggestions

Art

Plan an art tour of the public library, a local museum, company, or hospital known for its art collection. Note that this can be done from the library by viewing the online collections of museums. Tours can be in-person or virtual. Study the mathematical proportions, for example, the golden rectangle, in human figures or buildings or search for the patterns in fractals. A great introduction to these concepts combining mathematic, arts, and sports can be found in the short movie titled *Donald in Math-magic Land*. In one segment geometry is shown through the lens of playing pool (http://disney.wikia.com/wiki/Donald_in_Mathmagic_Land or on YouTube). See also Appendix A for a list of Art Museums and their websites.

Science

Science fairs and symposiums are another way to become involved with STEM activities. While science fairs can be held at any level, science symposiums are usually poster sessions held by high school classes such as biology. All involve research, statistics, charts or graphs, and a visual presentation. Mentors are always needed and librarians are in the perfect role to provide expert advice and leadership in designing research strategies and locating resources in print as well as electronic formats. Offer your services early in the year so there is plenty of time to plan. Does your school have an Environmental or Recycling Club? Help them to wade through government resources and the city website to find information on what is, and is not, being recycled. If they test water quality for pollutants in local streams, help them to report the data to the proper agency. In some cases, this may be a local park service. See Appendix B for a list of Science, Technology, and Natural History Museums and their websites.

Your work in the sciences does not need to be restricted to assistance for clubs. A program or series of programs on a regular basis might be helpful for patrons who aren't interested in competitions. National Public Radio, aka NPR, sponsors segments called Science Fridays (Sci-Fri). Their goal is to provide STEM programming each week. Would a session called "Delicious Smelling Chemistry" (May 23, 2011) draw some interest? Stories in the education section include a lesson plan with the grade level, time frame, tools needed, and experiment procedures. Their Sci-Fri Spoonful section asks a question, lists the standards met by the discussion, and provides videos and a transcript to answer questions such as "Why Does This Frog Glow?" (March 22, 2017). Just as they do you could schedule

Occupation	STEAM Categories
Actuary	Math
Airline Pilot	Math, Science, Computers
Anesthetist	Math, Science, Medicine, Health
Architect	Math, Science, Art
Artist	Art
Certified Nurse Anesthetist	Math, Science, Medicine, Health
Chemist	Science
Computer Programmer	Computer Science
Database Administrator	Computer Science
Engineer	Math, Science, Engineering—Biomedical, Chemical, Civil, Electrical, Mechanical
Forensic Scientist	Science, Art
Graphic Artist	Computers, Art
Materials Scientist	Engineering
Mathematician	Math
Mechanic	Math, Science
Nanotechnology	Science, Medicine
Optics Engineer	Math, Science
Painter	Art
Physician	Math, Science, Medicine, Health
Physician's Assistant	Math, Science, Medicine, Health
Physicist	Science
Research Analyst	Math, Computer Science
Researcher	Math, Science
Sculptor	Art
Software Engineer	Computer Science
Sports Therapy	Math, Science, Medicine, Health
Statistician	Math
Technical Illustrator	Math, Science, Art
Technical Writer	Math, Science
Technicians	Science, Art
Welding	Math, Science, Manufacturing (metals, coating)

Figure 5.3: STEAM Occupations

a recurring weekly science activity time, with challenges in various areas, for example, mapping the local watershed; designing a model for recycling in the locality (perhaps after a speaker from city government talks); or physics experiments (https://www.sciencefriday.com/educate/).

Math

A big push in today's schools is financial literacy. Create a program for students on creating a budget or lab report using a spreadsheet program such as Excel. If you don't feel comfortable undertaking this alone, check with math teachers or the trainer at the public library for assistance. Remember you don't have to be the expert in every field. Bring in financial analysts or investment counselors to explain how to build an investment portfolio, or how to save for college, or their first apartment. Teachers in business or economics courses would be happy to know about the sophisticated databases, company profiles, and indexes that the public library can supply to students. Invite an architect in to speak about the use of art and higher-level mathematics in designing buildings and ensuring that the structures are sound enough to withstand storms, wind, blizzards, or earthquakes. Bring in a caterer to discuss the formulas needed to supply food for a banquet of 300 people. For a list of additional STEAM career titles see Figure 5.3.

GROUP PROJECTS
Scout Projects

If you have made contact with local scouting organizations your library may well be the first stop for young adults working on the capstone project for scouts, namely, Eagle Project for Boy Scouts or young women working on a similar project for Girl Scouts or Girl Guides. These teens need to create an original project, figure out the costs, secure the materials, arrange for discounts or donations, and then complete the project. Projects can include building benches for the senior center or creating a garden for the high school courtyard.

Prior to this culminating activity, Scouts may be interested in some of the science speakers that might be part of public library programming. Consider partnering with troops to have their members be part of presentations on ecology, environment, or other topics related to their experiences. If you have science or natural history museums nearby check to see if materials can be loaned to teachers, libraries, or schools.

Team Competitions

Just as History Day and the National Spelling Bee have regional-, state-, and national-level competitions, so do the fields of mathematics and science. They are always searching for advisors and resources; many require multiple advisors for each team. Students can compete at the local level with the possibility of advancing to regional and national competitions. Librarians can always help to furnish resources for such teams or host meetings. Competitions foster critical thinking and utilize teamwork to solve challenges. Many are available for students from elementary through high school. Robotics team competitions are held for both middle school and high school students. The resource STEMfinity provides a listing of additional STEM competitions at https://www.stemfinity.com/STEM-Competitions. Some sample competitions are listed below; others can be located through the Resources page at the end of the chapter.

- The Intel International Science and Engineering Fair. http://www.intel.com/content/www/us/en/education/competitions/international-science-and-engineering-fair.html runs the ultimate in science fairs inviting winners of local, regional, state, and national competitions to their week-long fair. Those advancing to the international level are eligible for scholarship offers and network with CEOs of international companies in STEM and medical fields.

- The National Invention Convention. www.stemie.org. The STEMIE Coalition is open to young inventors in grades 4–12, they hold their annual convention in Washington, D.C. STEMIE stands for Science, Technology, Engineering, and Math linked to Invention and Entrepreneurship also using the acronym STEM+I+E.

- Math Counts. https://www.mathcounts.org/ fosters middle school competitions with groups that compete in math challenges.

- Odyssey of the Mind. http://odysseyofthemind.com/p/. Teams compete in various challenges where they must create hypotheses and test them using their problem-solving skills.

- Science Oympiad. https://www.soinc.org/. Teams of students compete in as many as 15 events in the fields of earth science, biology, chemistry, physics, engineering, etc.

Sports Statistics

Math teachers generate student interest in their subject by using real-world examples. Depending on the season, and the Olympics schedule, they may use math to teach percentages, such as earned run average, yards per carry, and other sports-related statistics. If you pull books for and add brackets to bulletin boards celebrating March Madness basketball tournaments or any other college sports championship playoff—Stanley Cup hockey tournaments, the World Series, and/or the Super Bowl— then you can build on that effort by working with a local math teacher or statistician to produce a program that will interest young adults. As Abigail Adams said, "Don't forget the ladies." Women's sports such as softball, field hockey, lacrosse, tennis, etc., allow the same opportunity for research and statistical analysis. A site popular with teachers has been the middle school site "Baseball Stats" from the National Council of Teachers of Mathematics (NCTM) (http://illuminations.nctm.org/uploaded Files/Content/Lessons/Resources/6–8/BaseballStats-AS-GameSheet .pdf). This could be useful for an enrichment activity or a summer reading game and with a few simple tweaks could be adapted for any sport that you choose. For additional information try the NCTM blog that is open to the public without membership at http://www.nctm.org/Publications/ Mathematics-Teacher/Blog/.

Imagine giving students a space to track statistics for school teams, colleges nearby, or any teams of interest. Kids could check trophy cases, record the awards, and track these achievements with charts and illustrations. What a great school service project this would be!

Try the annual Tech Challenge from the Tech Museum of Innovation in Mountain View, California. In 2017 youngsters were challenged to help explorers who needed to bypass ravines and cross a field of ice (https:// www.thetech.org/thetechchallenge). Also included for each year's challenge are lessons in math, science, engineering, and literacy under the title "The Tech Challenge 2017 Lessons" (https://www.thetech.org/49962813). By using the search button, you can also view lessons from the previous year's challenge.

Those of us who majored in subjects other than STEM fields need not be intimidated about creating programs, events, or collaborations in the fields of science, technology, engineering, or math. Figure 3.2 lists professional associations that you can contact to search for local experts who can help you. There is a wealth of online and print resources available to us as well as a number of experts in our cities and regions who would be willing to give us some free advice. This is one of the reasons that librarians are such valuable agents of change. We value learning enough to leave our comfort zones in order to assist our young adults. Carry on!

RESOURCES
General

HHMI Biointeractive. http://www.hhmi.org/biointeractive. Earth Science and Biology topics. Each topic starts with a summary including the number of lessons and time required. Useful for choosing programs and modifying it to fit your needs.

IXL Math. https://www.ixl.com/math/. Provides help for students with all levels and complexities of math and science problems.

Master's in Data Sciences. http://www.mastersindatascience.org/blog/the-ultimate-stem-guide-for-kids-239-cool-sites-about-science-technology-engineering-and-math/#STEM_Camps-3. There is a wealth of information here for you and your staff and for your patrons. Websites are divided into four groups called Fun for: elementary school kids, middle school kids, high school kids, and girls. Categories are further divided into websites, summer camps, apps and games, awards, career resources, science and technology contests, math contests, and grants and opportunities. In addition websites are listed for each state relating the data science specialties of universities within that state. After the list of colleges that are recruiting you will find listings such as Data Sciences in California (http://www.masters indatascience.org/schools/california/). Scroll down and you will also find a section detailing Data Scholarships. Be sure to share this information with your high school students, community members, and the local guidance counselors.

NASA for Educators. https://www.nasa.gov/audience/foreducators/index.html. Divided by grade levels, K-4, 5–8, 9–12, and Higher Education.

National Council of Teachers of Mathematics (NCTM). Illuminations. http://illuminations.nctm.org/Lessons-Activities.aspx. Lessons and ideas for programming from the math teachers national association.

National Public Radio. Their "Sci Fri Spoonful" section asks a question, lists the standards met by the discussion, and provides videos and a transcript to answer questions such as "Why Does This Frog Glow?" (March 22, 2017)

National Science Foundation Classroom Resources. https://www.nsf.gov/news/classroom/. A wealth of information in mathematics and the sciences. Websites are annotated and aimed at teachers but are useful for anyone.

Nature Magazine. Scitable. http://www.nature.com/scitable articles and information on science fields from *Nature Magazine.* Resource for AP Biology and undergraduate science programs.

NOVA Labs. http://www.pbs.org/wgbh/nova/labs/ PBS affiliated. This site contains information, videos, video quizzes, educators' guides, and meet the expert sections.

The Regents of the University of California. "How to Smile." 2015. https://www.howtosmile.org/. Provides many topics on fields including chemistry, mathematics, sciences, and the human body with activities that can be used at home, in libraries, and in school. Useful for programming and as a teacher resource and home school resource.

Science 360 Video. https://science360.gov/files/. Videos on a variety of science and mathematics topics and a section on K–12 education.

Teachers Try Science. http://www.teacherstryscience.org STEM lessons and experiments. Useful ideas for collaboration and library activities.

Techbridge. "Resources to Encourage Kids in Science, Technology, and Engineering." Two-page handout of activities useful for students. Websites are included for easy access to information on each one. http://www.techbridge girls.org/assets/files/what/family/Resources%20for%20Families.pdf

UBC Skylight: Science Centre for Teaching and Learning. https://sclt.science .ubc.ca. From the University of British Columbia. Includes lists of grants and useful for instructors and high school teachers of advanced science courses.

STEM Schools

STEM Schools Directory—United States. https://www.stemschool.com/schools. Contact information is given as well as lists of conferences, articles, and relevant book titles.

STEM Schools Directory—Canada. http://www.bbr-education.com/canada-high-schools/stem/the-stem-advantage-canadian-high-schools-invest-in-the-future/Lists. STEM schools in British Columbia with links to other provinces.

Competitions

Canada Virtual Science Fair (not current, from 2000 to 2014). http://www.virtual sciencefair.com. Contains pertinent information such a Project Components and a listing of previous science fair projects that can be searched.

National Science Teachers Association (NSTA). Angela Award for one female student (Grades 5–8). http://www.nsta.org/about/awards.aspx#angela

NSTA. *Science Fair Warm-Up: Learning the Practices of Scientists* books geared to grades 5–8, 7–10, 8–12, as well as a teacher's edition. http://www.nsta.org/store/product_detail.aspx?id=10.2505/9781936959235

NSTA Bright Schools Competition (Grades 6–8). http://brightschoolscompetition .org/

Toshiba and NSTA. *ExploraVision*. Competition for (Grades K–12). http://www .exploravision.org/

U.S. Army eCYBERMISSION Program (Grades 6–9). http://www.ecybermission .com/

REFERENCES

AASL American Association of School Librarians. *Standards for the 21st-Century Learner*. 2007. http://www.ala.org/aasl/guidelinesandstandards/learning standards/standards

Cunningham, Christine M., and Melissa Higgins. "Engineering FOR Everyone." *Educational Leadership* 72, no. 4 (2014): 42–47.

Donald in Mathmagic Land. Walt Disney Productions, 1959. http://disney.wikia .com/wiki/Donald_in_Mathmagic_Land

Leask, Amy. "20 American STEM Museums You Shouldn't Miss." In STEM, STEAM, and STREAM. Blog. May 6, 2013. http://enableeducation.com/20-american-stem-museums-you-shouldnt-miss/

Leask, Amy. "Canadian STEM Museums, Coast to Coast." In STEM, STEAM, and STREAM. Blog. May 1, 2013. http://enableeducation.com/canadian-stem-museums-coast-to-coast/

Lombardi, Abby. "Software Development Ranks as the Most In-Demand Skill for Tech Jobs." Wanted Analytics. September 5, 2013. https://www.wanted analytics.com/analysis/posts/software-development-ranks-as-the-most-in-demand-skill-for-tech-jobs

Master's in Data Science. "STEM Fun for Kids K-12." http://www.mastersinda-tascience.org/blog/the-ultimate-stem-guide-for-kids-239-cool-sites-about-science-technology-engineering-and-math/

Nager, Adams, David M. Hart, Stephen Ezell, and Robert D. Atkinson. "The Demographics of Innovation in the United States." Information Technology and Innovation Foundation. February 24, 2016. https://itif.org/publi cations/2016/02/24/demographics-innovation-united-states

November, Alan. "The Seven Questions Every New Teacher Should Be Able to Answer; #1." Blog. June 13, 2016 (reprinted December 30, 2016), p. 2. http://www.eschoolnews.com/2016/12/30/1-questions-new-teacher-answer/

Office of the President of the United States, President's Council of Advisors on Science and Technology. *Engage to Excel: Producing One Million Additional College Graduates with Degrees in Science, Technology, Engineering, and Mathematics,* 2012.

Rothwell, Jonathan. "The Need for More STEM Workers." *The Avenue. The Brookings Institution.* June 1, 2012. https://www.brookings.edu/blog/the-avenue/2012/06/01/the-need-for-more-stem-workers/

Search Institute. "40 Developmental Assets for Adolescents (Ages 12–18)." http://www.search-institute.org/system/files/40AssetsList.pdf

Skloot, Rebecca. *The Immortal Life of Henrietta Lacks.* New York: Broadway Books, 2011.

"The Tech Challenge 2017: Rock the Ravine." The Tech Museum of Innovation. 2017. https://www.thetech.org/thetechchallenge

Wright, Joshua. "America's Skilled Trades Dilemma: Shortages Loom as Most-In-Demand Group of Workers Ages." *Forbes.* March 7, 2013. https://www.forbes.com/sites/emsi/2013/03/07/americas-skilled-trades-dilemma-shortages-loom-as-most-in-demand-group-of-workers-ages/#fbb9b646397c

Xue, Yi, and Richard C. Larson. "STEM Crisis or STEM Surplus? Yes and Yes." U.S. Bureau of Labor Statistics, May 2015. https://www.bls.gov/opub/mlr/2015/article/stem-crisis-or-stem-surplus-yes-and-yes.htm

FURTHER READING

Alessio, Amy J., Katie LaMantia, and Emily Vinci. *A Year of Programs for Millennials and More.* Chicago: American Library Association, 2015.

American Association of School Librarians (AASL). "Learning Standards and Common Core State Standards Crosswalk." http://www.ala.org/aasl/standards/crosswalk

American Association of School Librarians (AASL). "Math Crosswalk." http://www.ala.org/aasl/sites/ala.org.aasl/files/content/guidelinesandstandards/commoncorecrosswalk/pdf/All_Math_Standards.pdf

American Association of School Librarians (AASL). "Reading Standards Literacy in Science/Technology." http://www.ala.org/aasl/sites/ala.org.aasl/files/content/guidelinesandstandards/commoncorecrosswalk/pdf/ReadingLitSciAllStandards.pdf

American Association of School Librarians (AASL). "Writing Standards for Literacy in History/Social Studies, Science, and Technical Subjects." http://www.ala.org/aasl/sites/ala.org.aasl/files/content/guidelinesandstandards/commoncorecrosswalk/pdf/WritingAllStandards.pdf

Cho, Janet H. "Lego Thrives with Social Media Strategy." *The Plain Dealer* (Cleveland, OH), September 9, 2016.

Freiberger, Marianne, and Rachel Thomas. *Maths Squared: 100 Concepts You Should Know*. London: Apple Press, 2016.

International Society for Technology in Education (ISTE). 2016 ISTE Standards for Students, International Society for Technology in Education, 2016. https://www.iste.org/

International Technology and Engineering Educators Association. *Advancing Excellence in Technological Literacy: Student Assessment, Professional Development and Program Standards*. Reston, VA: International Technology Education Association, 2007.

Jones, John I. "An Overview of Employment and Wages in Science, Technology, Engineering, and Math (STEM) Groups." In *Beyond the Numbers*. Bureau of Labor Statistics. https://www.bls.gov/opub/btn/volume-3/an-overview-of-employment.htm

Levy, Joel. *Chemistry in 100 Numbers: A Numerical Guide to Facts, Formulas, and Theories*. London: Apple Press, 2015.

National Council of Teachers of Mathematics (NCTM). *Principles and Standards for School Mathematics*. Reston, VA: National Council of Teachers of Mathematics, 2000.

National Council of Teachers of Mathematics (NCTM). *Principles and Standards for School Mathematics, Grades 6–8 Edition*. Reston, VA: National Council of Teachers of Mathematics, 2000.

National Council of Teachers of Mathematics (NCTM). *Principles and Standards for School Mathematics, Grades 9–12 Edition*. Reston, VA: National Council of Teachers of Mathematics, 2000.

National Science Teacher Association (NSTA). "Access the Next Generation Science Standards by Topic." ngss.nsta.org/AccessStandardsByTopic.aspx

National Science Teacher Association (NSTA). *Next Generation Science Standards: For States, By States*. Washington, DC: The National Academies Press, 2013. www.nextgenscience.org/get-to-know

National Science Teachers Association (NSTA). "Science Resources for Parents." 2017. http://www.nsta.org/parents/

Noyce Foundation. "Examining the Impact of After-School STEM Programs." July 2014. http://www.afterschoolalliance.org/ExaminingtheImpactofAfterschoolSTEMPrograms.pdf

Scieszka, Jon. *The Math Curse*. New York: Viking, 1995.

Shetterly, Margot Lee. *Hidden Figures*. New York: William Morrow, 2016.

Southorn, Graham, and Giles Spar. *Physics Squared: 100 Concepts You Should Know*. London: Apple Press, 2016.

Stewart, Becky. "Challenging Perceptions in the STEM Classroom." NSTA Blog. February 21, 2015.

Stuart, Colin. *Math in 100 Numbers: A Numerical Guide to Facts, Formulas, and Theories*. London: Apple Press, 2015 (Currently unavailable).

Stuart, Colin. *Physics in 100 Numbers: A Numerical Guide to Facts, Formulas, and Theories*. London: Apple Press, 2015.

Young Adult Library Services Association (YALSA). "STEM Resources." May 2017. http://wikis.ala.org/yalsa/index.php/STEM_Resources

6

◆ ◆ ◆

TECHNOLOGY

Since the introduction of the personal computer in schools and libraries some 30 years ago, there has been tremendous growth in technology use and in the sophistication of personal computing devices. Public libraries now serve as centers for Internet access, training, and devices for those who don't have them at home, and for all patrons. In schools, technology growth has led to a broadening of access to school resources and curriculum. Technology use has grown far beyond a few computers, labs for word processing and presentations, and online catalogs. Today the focus is on technology integration, teaching the use of tools in relation to projects, and units of study requiring their use. With the ubiquity of computers and other devices outside of school and the growth of computers in testing, one-to-one programs and other initiatives, computing technology is used in every discipline. Technology underpins the work of young people in all areas of their school career.

TECHNOLOGY USE IN SCHOOLS
AND LIBRARIES

In STEM, the "T" stands alone as a reminder that young people require awareness and adeptness in using technological tools, whether hardware, software, apps, or web resources. Often in middle schools, there may be a technology course or unit that focuses on basic functions of computing systems and the use of standard software applications such as word processing, spreadsheets, and presentation software. In high schools, there may be general courses in coding and programming, web design, robotics and game design, as well as more specialized courses in networking and Advanced Placement (AP) courses. In public libraries, training sessions in various applications and software products are offered to young people, and patrons of all ages. And in both settings, young people use computers for reading, research, discipline-specific online programs, and online coursework from early childhood onward.

Thus technology as a part of STEAM often involves the use of digital applications that allow students to create, whether in science, mathematics, or the arts. Within standards documents in all disciplines, there are statements acknowledging the need for students to use a variety of media sources and devices to explore the subject and to communicate what they know and can do. These statements have at their core the principle that technology use is a foundational element in all subject areas, integrated as appropriate to the discipline involved. Organizations like the International Technology and Engineering Educators Association (ITEEA) and the National Science Teachers Association (NSTA) have included technology in their standards for engineering and science. These standards now include technology as a part of applied science experiences, extending their use to experiments and research, as well as communication of the results of such work. These connections are discussed further in Chapter 10 of this book.

The International Society for Technology in Education (ISTE) has focused on how technology can improve and perhaps transform learning and teaching through supporting educators who align technology tools and usage to learning outcomes for young people (International Society for Technology in Education, Redefining Learning, 2). ISTE also has a set of standards specifically for identifying student outcomes in broadly based statements that can be used in any discipline. They are organized in groups that describe what each student should become: empowered learner, digital citizen, knowledge constructor, innovative designer, computational thinker, creative communicator, and global collaborator (International Society for Technology in Education, ISTE Standards).

Librarians and Technology Integration

What does this mean for school and young adult librarians? We can bridge the technology integration happening in the classroom to collaborative opportunities in the library, creating informal learning options for young people focused on meaningful technology experiences using technology tools. Just as librarians manage print materials and databases, our role includes helping young people become conversant in the use of technology for their "job" learning. Many may think young people already have such skills. Often their skills are focused on manipulating and moving through the entertainment side of the web. Gaming, online viewing, creating video clips, participating in social media are all viewed through a social and entertainment lens. Our role is to develop their skills in professional communication and creation for a school or job focus. These skills linked to STEAM activities in programs and curricula help young people understand how to navigate technology use in a work-related setting, fun for a purpose.

Help young people understand how to navigate technology use in a work-related setting, fun for a purpose.

In the end, technology integration isn't about the "stuff," but about the use and manipulation of tools and resources to build understanding and facilitate creation. Longtime educational technology advocate Alan November differentiates between learning about a device and moving toward designing learning experiences using the tool. He also warns against the proliferation of apps, without clearly defining how a specific app improves learning and achievement for young people. November's focus is designing experiences that cut to deeper thinking and learning. He poses six questions to turn the corner from using stuff to innovating learning; questions that can focus our programming efforts around issues of critical thinking rather than content knowledge. Since developing critical thinking is central to library instruction, all programming should help students increase their ability to use web resources analytically. The assignment should offer an opportunity for questioning skills that take young people in new directions as they explore. We should find ways for students to reflect on their learning and articulate how they approach problem-solving. With the use of web resources, the world is available, as is the possibility to make connections worldwide. We should be in the business of providing opportunities for students to do authentic activities that give young people a sense that their work is worthwhile, that the assignment challenges them to use their review of content and their skill set to create high-quality work (November). For the purposes of STEAM experiences in informal learning, think in terms of activities rather than assignments. Focus on the purposeful hands-on work that can be encouraged in both the library and in the classroom.

Partners in Technology Integration

Many schools have technology teachers or technology integrationists who work with teachers, or have students passing through their labs. These teachers are natural partners for librarians. Since they are often "singletons" in the school, like librarians, teaming together can provide both richer experiences for students and support for both programs. In the public library team up with technology trainers. Your skills support their work. Their experience and expertise can inform your work with young people. It is a win-win situation.

Depending on how technology as a school subject is handled in an individual school, the approach the librarian takes to collaborating on technology integration may take different forms. Is there a culture of sharing and collaboration already present? If not, that becomes the first step in partnering. Are there technology courses in place? Library programming could help raise interest and involvement through after-school and study hall activities. This is an opportunity for collaboration with those that teach the course. In the public library, are there already technology training sessions in place? How can the young adult librarian partner with those initiatives? What programming could be developed to support effective technology use by young people?

TECHNOLOGY INITIATIVES AND LIBRARY PROGRAMMING

While technology is integrated into every area of STEAM, the discussion in this chapter includes two areas that often find their home in more than one subject discipline, in makerspaces, and may be part of full curricula, or a long-term activity sessions: coding and gaming.

Coding is the use of computer languages or programming to create software, apps, and websites. In practice, coding can also be used to create interactive stories and games. Whether the term programming or coding is used, it is all within the realm of computational thinking, using computer science to solve problems and create solutions.

Gaming in education takes a number of different forms: cosplay, digital gaming, and virtual reality. Gaming can become part of any discipline, and may be approached in two ways. Young people can be involved in acting out or participating in a game-based activity or simulation. Secondly, they can be involved in the creation of games that support learning goals. Please note that a third area closely related to technology integration, robotics, is closely aligned with engineering and is featured in Chapter 7.

Coding

What Is Coding?

Coding is the latest iteration of computer training, focused on computational thinking rather than software and hardware use. Coding is the language that builds software, apps, and websites. Knowing how to code and understanding how coding works can help young people navigate devices and applications they encounter. How does it fit with STEAM? Coding can be used in all disciplines to create useful processes and practical applications in all areas. Giving students a taste of coding can spark an interest in delving further into coding to create and explore.

The challenge in providing experiences in coding is that it is not necessarily a skill that will come naturally to a librarian. Fortunately there are resources that provide guides and even full instructions for sessions on coding. Consider it an opportunity to learn along with the young people that you serve, knowing that modeling learning behaviors is a great opportunity in itself. Never fear, students who are interested can also become facilitators with you. It is a great opportunity to experiment and to grow along with students and patrons.

Coding Activities

Where to begin? Coding may not be in a librarian's roundhouse, so finding local contacts that can help is a great place to start. The first step can be seeking out young people who are keenly interested in developing apps or websites. Learning from young people not only benefits you as a professional, it validates the informal expertise young people have in the tech world. Teens can be featured experts in coding events, as can local business and academic professionals. Keep an open mind and be fearless in learning right along with your patrons and students.

Hour of Code

The Hour of Code initiative began in 2014 as an effort to interest young people in coding as a career, founded by companies and organizations like Microsoft, Apple, Amazon, Boys and Girls Clubs of America, and the College Board. As of 2016, the Hour of Code logged almost 190,000 events in more than 180 countries (Jacobson, 2016). This initiative can be a good place to build a knowledge base about coding.

The Hour of Code organization (https://code.org) uses the tagline "Anybody Can Learn" to promote coding for all, without barriers. Its mission is to improve student access to computer science skills and to promote the teaching of computer coding to students of all ages, genders, and backgrounds. Activities to introduce children and teens to computer programming can be found at https://code.org/learn. On this page you can tailor your results to level, age group, and the length of program you desire. After making a choice, you will find activities, tutorials, and lessons.

Hour of Code workshops seek to "demystify code" to show that anybody can learn the basics, and to broaden participation in the field of computer science. The emphasis is on creativity and the fun of coding, using games to teach computer skills, and problem-solving, and logical thinking. Games are designed to be flexible in terms of how people are grouped: individuals, pairs, and in larger groups, using projectors to share information if multiple computers are not available. Tutorials are geared to age groups, and tutorials are often connected to popular games such as *Star Wars* and *Minecraft*. The Hour of Code can also be linked to specific promotional events outside of its own event, such as Computer Science Education Week, held annually in December.

Beyond Hour of Code

The beauty of the Hour of Code site is that tutorials and support information are always available and can be used on any device and can be shared. Since students have no need to create accounts using the site is easier than many others, and does not require parental permission slips. The program can be exploratory, offered for students, parents, or community members. It can be a one-time program or resources can be used for a longer series of activities. Coding activities are also a natural match for makerspaces. Coding opportunities can be designed as group activities or set up as independent activity stations.

Coding by itself can be an intellectual enterprise. Pair coding with available robotics kits and devices that require programming to create activities and projects. Among many others, Raspberry Pi and Arduino provide simple circuits that require coding. Some examples of what could be done include:

- Use coding to program electronic dice to "roll" random numbers, setting up probability and statistics experiments.
- Make a website. Use it to post information about coding and making.
- Design a mapping activity to outline an imaginary world, using distance measurements and other mathematical concepts.
- Develop an app around a topic that help students understand science or math concepts, or around an identified need in the school, library, or community.

Resources for Coding

Alice. http://www.alice.org
> Use Alice to create animations, interactive games, or videos.

Code Academy. https://www.codecademy.com
> A free sign-up provides access to tutorials for a variety of programming languages. The "For Education" page includes teacher training, classroom resources, and information for setting up student accounts.

Code.org. https://code.org. Use the https://code.org/learn for programming ideas

Computer Science Education Week. https://csedweek.org
> The Educator Resources section has resources to learn more about coding, the Hour of Code, and third-party products to use.

Khan Academy. https://www.khanacademy.org/computing/computer-progra mming
> There is an extensive section on this website that includes coding, creating graphics, web design, simulations, visualization, and more with videos and coursework. Individual accounts can be set up, and there is an option for teachers to set up class groups.

Libraries Ready to Code. American Library Association. Dec 8, 2016. https:// www.youtube.com/watch?v=vFBZz9_TVXc

ScratchEd. http://scratched.gse.harvard.edu
> Scratch is a free coding language using blocks to interlock to create multimedia projects. This site supports the Scratch programming language popular in educational settings, with many resources. Among them:

>> Brennan, Karen, Christian Balch, and Michelle Chung. (No Date) *Creative Computing*. Harvard Graduate Program of Education. http://scratched .gse.harvard.edu/guide/files/CreativeComputing20141015.pdf
>> Learn how to use Scratch for creating interactive stories, games, and animations.
>> Creative Computing. http://scratched.gse.harvard.edu/guide/index.html
>> This site focuses on building creativity for 8–16-year-olds using Scratch. There is also a downloadable guide for curriculum and general activities for educators in the classroom and in libraries, including handouts and

procedures. There is also a workbook version, and editable versions for non-commercial use under a Creative Commons license.

Trucano, Michael. "Learning to Code vs. Coding to Learn." February 9, 2016. http://www.techlearning.com/blogentry/10318
This thought piece on the pros and cons of coding in schools is useful in determining how a school might best implement coding in the curriculum, and what any organization should consider before implementation.

Gaming

People of all ages love gaming. Games are fun. They are usually played with friends. Because teens view games as "fun," they are willing to take risks, to fail, and to try again until they either win or advance to the next level or master the game (Plass, 4). Yet games are often complex, require practice, and promote problem-solving. So, this willingness to risk failure and persevere until success is attained is at the heart of the scientific method and the design process. These aptitudes also help build for success in the real world as inventors, scientists, and entrepreneurs. Gaming comes in many forms and formats, all of which can work in any library setting: traditional board games, online gaming, cosplay (costume play), or virtual reality. They are wildly popular with pre-teens and teens.

Gaming in the library facilitates flexibility and creative thinking with low risk and high engagement. Games allow you to foster STEAM habits of mind and skills. Combined with collaboratively planned gaming involving a classroom teacher and the school librarian, gaming can marry high student engagement with content understanding.

Cosplay and the Arts

Since the days of Dungeons and Dragons tournaments, young gamers have enjoyed taking on roles and acting out the parts. Think of the library as a place for gamers to convene, with that great display of books matching the activity to lure them into reading as well. Whether Renaissance dress, steam punk dress, or the latest space costuming, imagine the fun of promoting these activities via special displays. Activities can allow students to design, create, or enhance costuming with some expert advice. Bring in experts from a fabric store to help students to select fabric or accessories. Ask the curator of the local art museum to discuss the intricacies of chain mail and armor. Perhaps the historical society could provide a speaker to talk about local inventors of the last century. Check with the high school drama coach and Thespian troupe, or the local community theater director to make a presentation or share props with you. If there is a professional theater company nearby, check to see if they have an educational programming department or speakers bureau, and take advantage of the offerings.

Escape Rooms

Lock-ins have proven successful in schools, recreation (rec) centers, and libraries; but they don't have to involve an overnight stay to be successful. Consider an escape room activity that can be done after hours. Youth Librarian Karissa Alcox of Fort Erie Public Library (Ontario, Canada) details the steps she took to create teams for a short (15-minute) game that allowed teams to follow the clues provided, to collaborate with their team, and to determine how to escape the room in the time allotted (O'Reilly, 2016, 14). Alcox explains the storyline she created, and how she promoted the activity using social media as well as flyers to promote her escape room program. She planned a community event, open to the public, which provided 52 people with the chance to try an escape room fantasy that took only 15 minutes. She and other librarians have used experts in escape room activities who can help create an event with just a few props (O'Reilly, 2016, 17). Kits are available and can be purchased from Breakout EDU, but if budget is an issue simple props can be created or repurposed from items in the library or from donations. Props can include locks and keys, passwords, treasure chests, puzzles, whiteboards, blacklight flashlights and curtains, a countdown clock or timer, and books with invisible ink. Start planning the activity by creating a chart, filling in the details as you flesh out your activity. This will help manage the activity and ensure that all elements of the game are covered, as in the example chart (Figure 6.1). Figure 6.2 shows a blank chart that can be used for any gaming situation.

Game Name	Island Escape
Time Limit	15 minutes to 2 hours
Group Size	2–6 players
Clue Levels	Easy to Advanced
Ground Rules	Areas off limits Time allotted
Red Herring Clues	3–15 (list them below)

Figure 6.1: Game Planning Chart Sample

Game Name	
Time Limit(s)	
Group Size	
Clue Levels	
Ground Rules	
Red Herring Clues	

Figure 6.2: Game Planning Chart

From *Full STEAM Ahead: Science, Technology, Engineering, Art, and Mathematics in Library Programs and Collections* by Cherie P. Pandora and Kathy Fredrick. Santa Barbara, CA: Libraries Unlimited. Copyright © 2017.

An escape room activity could tie in to events like the American Library Association-sponsored Teen Tech Week in March or the International Games Day in November (O'Reilly, 2016, 17).

Games and Critical Thinking

Social games where another player or team is required can help participants to plan ahead, to predict opponents' next moves, and to collaborate with other team members. These games help young patrons build critical thinking skills that will be useful in STEAM fields and in the workplace for any career. Simple board games such as checkers require an analysis of the other player's move while more complex games such as Risk, Stratego, and chess require close attention and the ability to make predictions about behavior of other players. The same type of strategies can be found in online and computer gaming programs. These may already be in the library collection or linked on the library website. Surveying how teens access games on computers in their downtime can help determine which ones are the most popular.

Game-Based Learning

There has been a sharp increase in the use of game-based learning in schools. Project Tomorrow's annual Speak Up survey reports that in 2012, 30 percent of teachers used games in the classroom. By 2015 the number rose to 48 percent (Evans). What elements make gaming attractive in working with young people? Beyond the fun factor, young people find self-efficacy, a sense they can succeed, in working through problem-solving situations in a game. Games also bring a high level of engagement if used in group settings, teamwork, and communication skills. Involving young people in groups also requires them to analyze, create, and develop a deeper understanding of the situations presented by games.

What role can libraries play in game-based learning? One role is in the provision of games in the library collection. This extends to curation of online game sites that support learning goals in STEAM and throughout the curriculum. In the library program, gaming can be linked to classroom activities and it can be part of informal learning experiences in the library. Whether a club activity or a series of events, gaming can be sure draw for young people.

Tasha Squires, teacher/librarian at O'Neill Middle School in Downers Grove, IL, created a literacy game called *Conquest of the Realm* (CotR) as a part of her curricular unit. This "gamification" of the process, as she called it, allowed students to work within teams, to collaborate, to solve problems, and, most importantly, to make their own decisions as to which activities (called challenges) they would pursue. Tools included gaming challenge cards and a billboard. Placing students into "houses" is a long-held practice in middle schools and students are familiar with the concept through the Harry Potter books and movies. Students enjoyed competing against each other, and against other houses, for points. Participation was voluntary but students often worked online outside of class hours as well as face-to-face. Squires found that interest in reading books and writing book reviews lasted long after the four-week unit was completed and students enjoyed creating book trailers (Squires, 2016, 20).

An added benefit of gaming is that a new audience may be drawn into the library where they are often prompted to explore other services and programs. In 2010 San Jose State University began a study of librarian versus teen views of their libraries. Each group was asked to create a two- to three-minute video tours of their libraries. Teen videos were then compared to those produced by librarians. One of the conclusions was that librarians focused on resources and services while teens focused on what they did in the library. Important to them were places to work alone and in groups and places to lounge and chat with friends (Kuhlmann, 2014, 24–28). Thus part of what can happen in libraries is the creation of spaces and resources that draw young people in, including gaming spaces and options.

Digital Gaming

With the increase in computing devices in libraries and schools, digital gaming can be supported to take advantage of the interest young people take in gaming. In schools, simulations have been used for learning experiences from math scenarios to science dissection and more. Gaming offers another possibility for students to problem-solve within a structured environment that sets challenges for them. A short list of possibilities, some of which now have educational versions:

- *Angry Birds*: mathematics applications around angles and physics concepts
- GlassLab Games's *Ratio Rancher*: mathematics, grades 6–12
- *MineCraftEDU*: designing, building, and mathematical concepts like fractions
- *SimCityEDU*: environment, science concepts
- *World of Warcraft*: ecology, climate, mathematics (volume, area, perimeter)

One of the newest developments in gaming brings a focus on virtual reality through the use of devices like Oculus Rift and Google Cardboard. These devices immerse a user in a 4-D environment using headgear and software that gives the user the feeling of being immersed in a simulated world. At this writing, VR tools for schools and libraries are in its infancy, and technology educators are exploring the full role of these tools in the school program. Including VR in a makerspace or after-school activity would make a good place to experiment with this technology.

Game Creation

To this point, the gaming discussion has been around playing games in all their formats. A further step in STEAM activities is game creation, whether on paper or digital. On paper, it is an opportunity to integrate art into a design experience. Digitally, it can involve coding, and lead young people to develop apps to match their interests. This can be a point where the library can involve young people in developing apps that may benefit the library or school community.

What benefits will young people gain? They will explore the parameters of gaming and learn protocols. Since games at their core run by a unique set of rules, there can be discussions of what would apply in a given game scenario. They would also need to determine what content-specific concepts would be at play in their game, and the impact that those scientific, mathematical, or artistic concepts would have on the functions of the game.

How can a library manage game creation opportunities? One option is to host a game jam, where teams are challenged to design a game in a short time (Farber). These events are often organized around a theme, such as using Earth Day as an opportunity to design games around ecological concerns. In conjunction with a subject discipline, this could be a type of culminating project that could be collaboratively managed by a librarian and teacher. In a public library setting, there could be periodic jams aligned to local initiatives.

Gamification

Another angle on gaming is the concept of gamification. This concept is not gaming per se. Gamification involves taking elements of gaming, such as accumulating points, rewards, and role-playing, into daily classroom or library activities. It is not implementing a game, but using elements to draw young people into a more interactive environment. While this would not involve specific projects, it does offer a direction for organizing and realigning processes to be more patron/student friendly. An example of this is the way Le Sébastien High School in Quebec teaches physics. Students are assigned roles as mages, warriors, and heroes in an application called Classcraft (Johnson, 2014, 39). Imagine how engagement would grow for students who do not necessarily look at physics as an involving experience.

In practical terms, game creation can begin with paper-pencil board game design. As students and patrons become more comfortable with coding, they can design and develop online games. Drawing on local expertise, from community members to tech-savvy students, is a great place to begin.

Resources for Game-Based Learning

American Library Association. Games and Gaming Round Table (GameRT). http://www.ala.org/gamert/
   The site includes game information, library resources, and discussion groups (some only available through committee membership).
Common Sense Media. https://www.commonsensemedia.org
   This site reviews games (and other technologies) for their value and use with young people, searchable by age, platform, discipline, skills, and purpose. The educator section includes lesson plan resources and other learning options.
Gamifi-ed Wiki. www.gamifi-ed.wikispaces.com
   Students and educators collaborate on this wiki to identify games with curricular content and educational merit following an established rubric for inclusion,
Google Cardboard. https://vr.google.com/cardboard/
   Google's entry into the virtual reality field has the advantage of being reasonably priced, with educational resources identified by Google on this site.
Gray, Dave, Sunny Brown, James Macanufo. *Gamestorming: A Playbook for Innovators, Rulebreakers, and Changemakers*. Sebastopal, CA: O'Reilly Media, 2010.
International Game Day @ Your Library. http://igd.ala.org
   This site includes resources for tabletop, board, and online games to be celebrated on International Game Day in November of each year.
International Society for Technology in Education. "Games and Simulations Network." http://connect.iste.org/communities/community-home?CommunityKey=95d9a716-6cdf-4599-a2f4-03a240ab752d
   This online community requires ISTE membership to access resources and discussions.
Oculus Rift. https://www.oculus.com
   Examine this site to learn more about virtual reality devices in this newly expanding field. Educational resources are also provided.

Pike, Jim and Milo Lam. "Getting Started with Minecraft in the Classroom." Common Sense Media. Blog. January 15, 2016. https://www.commonsense. org/education/blog/getting-started-with-minecraft-in-the-classroom
Pike and Lam cover demonstrating *Minecraft*, developing lessons, and integrating the game into project-based lessons, as well as determining student engagement and common commands.

Schrock, Kathy. Augmented and Virtual Reality in the Classroom. http:// www.schrockguide.net/augmented-reality.html
Find a plethora of resources from activities to apps to how-to guides.

Technology use involves a set of skills and abilities for young people to master. These skills can be approached in all areas of STEAM. In a best-case scenario, technology tools give young people a many avenues for learning, communicating, and creating with a sense of self-efficacy that will enhance their experiences in their school career and beyond. Libraries are perfectly aligned to go beyond collection building and curation of online resources to involve young people in activities that will spur academic growth and personal satisfaction.

REFERENCES

Evans, Julie. "Speak Up 2015 National Results: From Print to Pixel." http://www.to morrow.org/speakup/from-print-to-pixel-may-2016.html

Farber, Matthew. "Students as Designers: Game Jams!" July 3, 2015. https://www.ed utopia.org/blgo/students-as-designers-game-jams-matthew-farber

International Society for Technology in Education. "ISTE Standards for Students." 2016. https://www.iste.org/standards/standards/for-students-2016

International Society for Technology in Education. "Redefining Learning in a Technology-Driven World." June 2016. http://www.iste.org/docs/ Standards-Resources/iste-standards_students-2016_research-validity-report_final.pdf?sfvrsn=0.0680021527232122

Jacobson, Linda. "Coding's Finest Hour." *School Library Journal* 62, no.1 (January 2016): 11.

Johnson, Larry, Samantha Adams Becker, V. Estrada, and A. Freeman. *NCM Horizon Report: 2014 K-12 Edition*. Austin, TX: The New Media Consortium, 2014. https://www.nmc.org/publication/nmc-horizon-report-2014-k-12-edition

Kuhlmann, L. Meghann, Denise Agosto, Jonathan Pacheco Bell, and Anthony Bernier. "Learning from Librarians and Teens about YA Library Spaces." *Public Libraries* 53, no. 3 (May/June 2014): 24–28. http://publiclibrariesonl ine.org/2014/07/learning-from-librarians-and-teens-about-ya-library-spaces

November, Alan. "Clearing the Confusion between Technology Rich and Innovative Poor: Six Questions." November Learning. January 12, 2015. http:// novemberlearning.com/assets/ClearingtheConfusionbetweenTechnology RichandInnovativePoorSixQuestions.pdf

O'Reilly, Katie. "Libraries on Lockdown: Escape Rooms, a Breakout Trend In Youth Programming." *American Libraries*, 47, no. 9/10 (September/October 2016): 14–17.

Plass, Jan. "Why Games and Learning." Institute of Play.org. http://www.institute ofplay.org/about/context/why-games-learning/

Squires, Tasha. "Engaging Students through Gamification." *American Libraries* 47, no. 3/4 (March/April 2016): 20.

FURTHER READING

Adams Becker, S., A. Freeman, C. Giesinger Hall, M. Cummins, and B. Yuhnke. *NMC/CoSN Horizon Report: 2016 K-12 Edition*. Austin, Texas: The New Media Consortium, 2016. https://www.nmc.org/publication/nmc-cosn-horizon-report-2016-k-12-edition

Alvarez, Vivian. "Engaging Students in the Library through Tabletop Gaming." *Knowledge Quest* 45, no. 4 (March/April 2017): 40–48.

Elkins, Aaron J. "Let's Play! Why School Librarians Should Embrace Gaming in the Library." *Knowledge Quest* 43, no. 5 (May/June 2015): 59–63.

Institute of Play. "Why Games & Learning." http://www.instituteofplay.org/about/context/why-games-learning

Jacobsen, Linda. "The Surprising Impact of Brain Games on Learning." *School Library Journal*. October 4, 2016. http://www.slj.com/2016/10/industry-news/the-surprising-impact-of-brain-games-on-learning/

Kim, Bohyun. "Gamification As a Tool." *American Libraries* 46, no. 3/4 (March/April 2014): 26.

Martin, Crystle. "Expressing Youth Voice Through Video Games and Coding." *Knowledge Quest* 45, no. 4 (March/April 2017): 50–57.

Martin, Crystle. "Libraries as Facilitators of Coding for All." *Knowledge Quest* 45, no. 3 (January/February 2017): 47–53.

Moorefield-Lang, Heather. "Libraries and the Rift: Oculus Rift and 4D Devices in Libraries." *Knowledge Quest* 46, no. 5 (May/June 2015): 76–77.

Neason, Alexandria. "What Does It Mean To Have Your Whole Middle School Curriculum Designed Around Gaming?" *Hechinger Report*. August 13, 2015 http://hechingerreport.org/what-does-it-mean-to-have-your-whole-middle-school-curriculum-designed-around-games/

"Playing It Smart; How and Why Game-Based Learning Delivers Academic Results." *District Administration* 52, no. 5 (May 2016): 62–66.

Project Tomorrow. "Digital Teachers, Digital Principals; Transforming the Ways We Engage Students." October 3, 2014. http://www.tomorrow.org/speakup/downloads/PROJECT-TOMORROW-10-3-14.pdf

Pun, Raymond. "Winning Ways to Gamify Your Library Services." *Computers In Libraries* 36, no. 9 (November 2016): 12–15

Sansing, Chad. "Of Coding and Compassion." *School Library Journal* 42, no. 4 (April 2016): 28–30. http://www.slj.com/2016/04/technology/of-coding-and-compassion

Wilson, Michelle. "Coding: What's In It for the Library." Blog. December 29, 2016. http://knowledgequest.aasl.org/coding-whats-library/

7

◆ ◆ ◆

ENGINEERING FOR YOUNG ADULTS

In the past decade, a new "science" has been emerging in schools across the United States. Engineering, with its focus on problem-solving and design process is showing up in high school coursework, in middle school units of study, and in science-related special events in schools and communities. It complements science instruction, and as an applied science, offers hands-on experiences. Scientists work to understand natural phenomena. Engineers work to design a solution to a problem, often grounded in scientific and mathematical principles. Scientists focus on the natural world. Engineers focus on the built world—man-made products. Scientists collect data to understand natural phenomena. Engineers develop processes, products, and systems. Science learning focuses on universal knowledge; engineering focuses on specific situational learning. Together they provide both conceptual and hands-on experiences to deepen understanding and promote learning.

Engineering as a part of K–12 education can be a vehicle for problem-based learning (aka PBL) that introduces young people to concepts and principles they learn in science and mathematics, and that they then apply using technology skills. This approach can build valuable skills around teamwork and communication, as well as fostering creativity and imagination. Using a design process helps students imagine solutions to practical problems and then test out the pros and cons of such solutions. It can help students see the complexity in solving issues, and help them understand the need to apply creativity to the solution. Once there is a possible solution in play, basic concepts of engineering can be introduced to test solutions. Students then have a vested interest by sharing their own ideas that allows educators to engage them more fully in instruction. The focus on application and integration builds achievement and growth for students, giving them a level of preparedness for the STEM workplace.

ENGINEERING COURSE OFFERINGS IN SECONDARY SCHOOLS

In many states, there is not a requirement for engineering as a K–12 discipline, which can be a roadblock to developing programs in schools. As new standards documents are being developed in mathematics, and especially in the sciences, engineering is being seen as a practical approach to using conceptual knowledge. Thus, engineering often becomes a part of existing science classes, as a unit of study or exploratory experience. Statistics on the establishment of K–12 engineering programs are limited. This being said, the move to offer engineering is growing. Engineering is Elementary (EiE) and Project Lead the Way (PLTW) are two organizations that have collected information about enrollment in secondary education programs and the use of their curricular materials. As of 2013, EiE estimated that 52,000 teachers and 4.1 million students had used their curriculum. PLTW reported in 2014 that 5,500 schools offer a minimum of one program yearly, with estimates of 400,000 and 500,000 students participating each year (Honey, 2014).

What got the K–12 engineering initiative started? The state of Massachusetts was an early adopter of engineering in K–12 education, where the state Board of Education adopted new technology and engineering standards on December 20, 2000, followed by the framework document in 2001. They also included these standards in the state's assessment program (Massachusetts Department of Education, 2000, 20). The National Science Foundation (NSF) then began to fund K–12 engineering curriculum development and professional development. In 2004 the NSF established the National Center for Technological Literacy to spearhead these efforts (Miaoulis, 2014, 28–29). The work by the NSF led to the development of the Next Generation Science Standards, which included engineering as a part of the science curriculum, and have been widely adopted.

Elements of Engineering Instruction for Middle and High School Students

What does engineering mean in the context of young people in middle and high school grades? To begin, we need a basic definition of engineering, as it would apply to secondary schools and young adults. Engineering is an applied science, a practical application of scientific principles using a design process and step-by-step practice to develop solutions to identified problems (National Research Council, 2012, 11). This broad definition is a good jumping-off point for developing activities, events, and programs to engage and challenge students in engineering work. On a practical level, activities around designing and building are natural draws for students. The best of engineering programs and projects are hands-on and highly engaging for all students.

Goals of Engineering Education

What are the goals of engineering education for young people? Why are schools building engineering courses and units of engineering study in science classes? These reasons may speak to their importance in the K–12 curriculum:

- Helping students learn skills like communication, teamwork, and creativity—mirroring the approach of arts and humanities that have traditionally used this focus
- Helping students learn a systematic method for identifying and solving problems
- Preparing students to be aware and connect with STEM skills as they enter adulthood
- Motivating students to engage in science and math practical activities
- Helping students see connections between the science, math, and engineering concepts they experience in school as tools to address societal issues today and in the future
- Exposing and preparing students for STEM career possibilities (Moore, 41–42)

DESIGN PROCESS

Key to instruction in engineering concepts is the engineering design process. These steps can be used by groups of students to guide their work in solving problems. As can be seen in Figure 7.1, this cyclical process

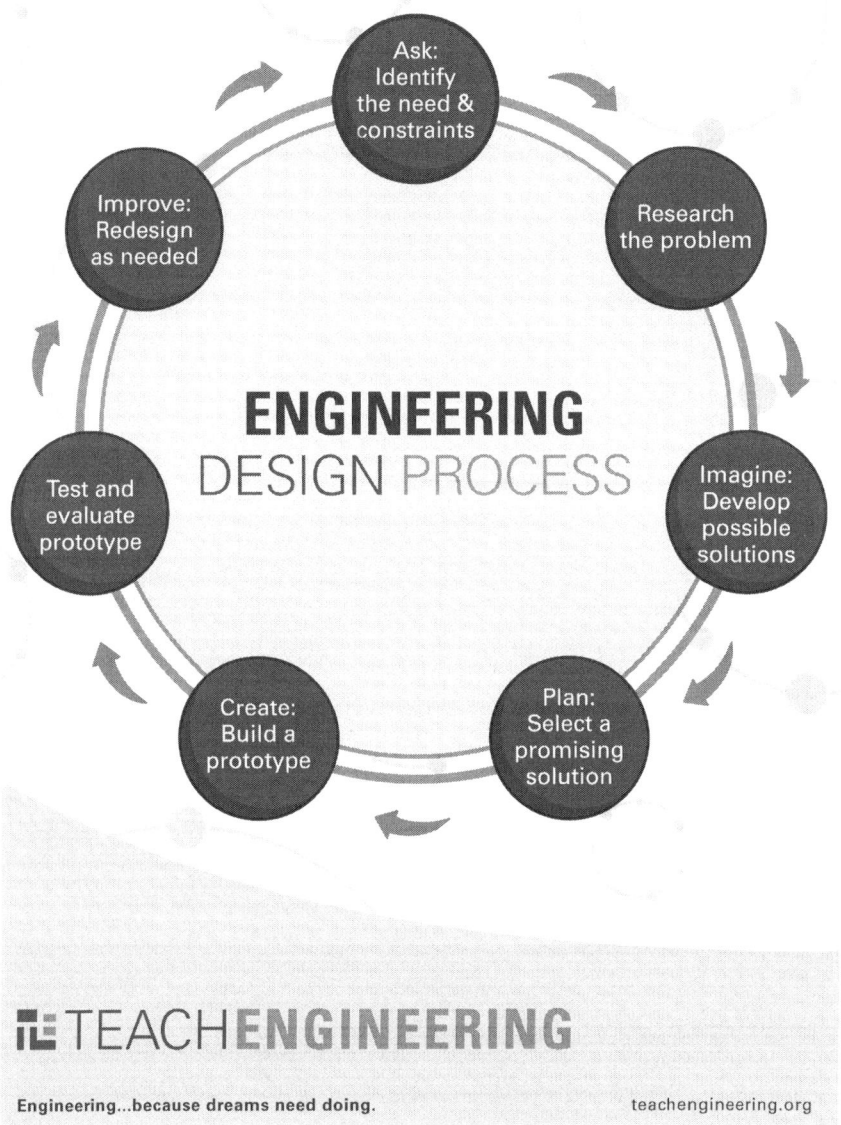

Figure 7.1: Design Process Chart

©2014 Kaitlyn O'Toole, TeachEngineering Digital Library. Used with permission. https://www.teachengineering.org/k12engineering/designprocess

Ask: Identify the Need & Constraints

Engineers ask critical questions about what they want to create, whether it be a skyscraper, amusement park ride, bicycle or smartphone. These questions include: What is the problem to solve? What do we want to design? Who is it for? What do we want to accomplish? What are the project requirements? What are the limitations? What is our goal?

Research the Problem

This includes talking to people from many different backgrounds and specialties to assist with researching what products or solutions already exist, or what technologies might be adaptable to your needs.

Imagine: Develop Possible Solutions

You work with a team to brainstorm ideas and develop as many solutions as possible. This is the time to encourage wild ideas and defer judgment! Build on the ideas of others! Stay focused on topic, and have one conversation at a time! Remember: good design is all about teamwork!

Plan: Select a Promising Solution

For many teams this is the hardest step! Revisit the needs, constraints and research from the earlier steps, compare your best ideas, select one solution and make a plan to move forward with it.

Create: Build a Prototype

Building a prototype makes your ideas real! These early versions of the design solution help your team verify whether the design meets the original challenge objectives. Push yourself for creativity, imagination and excellence in design.

Test and Evaluate Prototype

Does it work? Does it solve the need? Analyze and talk about what works, what doesn't and what could be improved.

Improve: Redesign as Needed

Discuss how you could improve your solution. Make revisions. Draw new designs. Iterate your design to make your product the best it can be.

And now, REPEAT!

Figure 7.1 (*continued*)

From *Full STEAM Ahead: Science, Technology, Engineering, Art, and Mathematics in Library Programs and Collections* by Cherie P. Pandora and Kathy Fredrick. Santa Barbara, CA: Libraries Unlimited. Copyright © 2017.

doesn't need to end with a final product, as its continuing use can make improvements in the work. In our classrooms, students have deadlines and handing in a project usually means the work is finished. In engineering, there is a continuing process focused on making improvements, which can be a good career skill and a key to lifelong learning. Note that this design process bears some commonality with the library research process, and with the scientific method.

Engineering is a process involving teamwork, another critical skill for the workplace and a key to continued growth as a learner. The ability to share information, participate in group discussion(s) and work, and use creativity in this setting are valuable skills to carry from school to other settings. While they need to be responsible on an individual level for understanding content knowledge, the experience of working and communicating together is equally important. Team activities where each participant has a specific role give young people experiences to carry with them into the work world. This is reflected in the most recent standards for science and mathematics. (See Chapter 10 for more information on standards.)

Engineering encourages experimentation based on perceived solutions that are then tested out in hands-on activities. While some may see this as trial and error, there is more to it than just "messing around" in standard engineering practice. In practice, the hands-on activity is closely linked to the proposed model developed prior to experimentation. This helps students to reflect and justify their actions. The final steps they undertake do not end with a successful solution; the final steps are reviewing initial goals set for the process, and suggesting possible changes to improve the process and/or product.

Reflection in Education

The final steps they undertake do not end with a successful solution; the final steps are reviewing initial goals set for the process, and suggesting possible changes to improve the process and/or product.

While engineering instruction can be part of a classroom curriculum, the library can also encourage students to explore engineering in hands-on activities. In school libraries, classes benefit from the teacher-librarian team presenting and coaching students through projects that involve library research and brainstorming. As libraries of all kinds develop makerspaces, the focus on hands-on projects and learning in an informal setting lends itself to engineering design activities. The opportunities for informal learning help encourage students who may be reluctant to jump into formal courses. Programs and activity centers can foster learning about the design process and creative thinking about real-world problems. What might the activities look like in a library setting?

- As students study the environment and the impact of recycling, have them study the pros and cons of a variety of packing materials (e.g., plastic bubble wrap, Styrofoam peanuts, shredded paper) to determine what would have the least impact environmentally while still protecting the resource they contain. Using an engineering process model, they examine the criteria to use in developing packing materials and the constraints in being able to successfully complete their activity.

- Since engineering involves much more than building things, projects that analyze processes or systems to evaluate their effectiveness are also fair game. This could take the form of analyzing how to separate an ore (e.g., iron) from sand or dirt that surrounds it, to a specific purity level. Participants could identify the best process to create a chemical reaction, or an energy-efficient way of removing leaves in the fall.

- Teens and tweens could be challenged to solve a local community issue like improving a park area, using the design model to work through the issues involved. They should explain why their plan is workable, addressing the impact on community members and on the environment. Solutions could address making the space accessible to special needs children, providing adaptations for the elderly, and implementing green solutions for waste.

- Provide a design brief to students, then have them create their own as a part of a team. The design brief is a brief (usually one-page) overview of the plan for a project. It would include statements on the problem under consideration; the audience for the product or solution; the materials and resources available; the constraints of the project; and the criteria for evaluation. While examples can be shared, student engagement may be stronger when they choose their own issue to examine. Two links that provide more information on the brief are:

  - Science Buddies sample worksheet. http://www.sciencebud dies.org/engineering-design-process/engineering-design-brief-worksheet.pdf

  - Project Lead the Way Design Brief. http://docslide.net/docu ments/design-brief-2011-project-lead-the-way-incintroduc tion.html

- Single-session activities could include building activities, using found materials or items like Legos or other commercial available building materials. A challenge could be set for the start of the session, like a structure that could bear a certain amount of weight, or of a certain height or other specified dimensions.

Parallels between the Engineering Design Process and the Library Research Model

What commonalities exist between school library curricula and classroom engineering curricula? Both focus on an iterative process. Librarians use a research model. While there are many iterations of the research model, basic steps include:

- Determining the question or problem to be addressed.
- Locating available resources.
- Evaluating how information meets the information need.
- Synthesizing information into a product.
- Evaluating the process and product, revising as needed.

The design process is integral to developing engineering products. In engineering, the design process includes:

- Determining the problem to be solved.
- Identifying factors to be considered in choosing the best solution using specific criteria for evaluation.
- Designing solutions: multiple ideas of possible solutions are evaluated to determine which will be best in a given situation.
- Testing of possible solutions.
- Refine promising solutions.
- Evaluating effectiveness and communicating results.

Both library and engineering processes allow for a nonlinear approach, where steps can be repeated as needed with a focus on solving the problem at hand. Both models are familiar to a scientist, who employs a scientific method that approximates these processes. This is all to say that features of engineering—in the use of a design process—can inform the creation of activities and programs that draw young people in to the process. Unlike traditional classroom approaches where lecture might be the norm, a classroom built around the engineering design process would include stations where students experiment and build and try out their ideas and creativity. This can easily extend to activities in the library, where the less formal setting will lend itself to students working on projects.

The hands-on approach is a benefit for students at all levels. If an introductory engineering course were developed early in the secondary curriculum, students could then schedule math and science course that build toward their needs in post-secondary education, whether technical school or university. This introductory course might be the vehicle to introduce students to systems and design thinking that contributes to problem-solving in all areas of their studies.

Whether developing engineering curricula or more informal activities in the library, consider how to make the offerings accessible and accommodating to all young people. The nature of projects, based on real-world issues, is a natural draw. Inviting engineers in to talk about their work and have conversations about their role will help students determine if this is a field for them going forward. The reality of engineering is that there is rarely just one solution. This opens doors for students to explore creatively and adjust their work when one possible solution does not work. This is also a collaborative process where all are welcome and working together is a must.

Engineering Strengths

The reality of engineering is that there is rarely just one solution. This opens doors for students to explore creatively and adjust their work when one possible solution does not work. This is also a collaborative process where all are welcome and working together is a must.

Engineering Design Challenge Ideas and Materials

As a librarian and educator, it is key to develop creativity and perseverance in young people, so activities should include both systematic analysis and creative thinking. It is critical to understand how all parts interact and interrelate, keeping variables in mind in drawing conclusions and developing solutions. There are also opportunities for discussion and research on social and environmental issues that could impact design. Some of the tools that can be helpful in aiding project design include:

- Testing and measurement items such as thermometers, oscilloscopes, and indicators.
- Software for data acquisition, starting with basic spreadsheet software.
- Computational and visualization tools, such as graphing calculators, computer-aided design (CAD) programs.
- Internet resources such as mapping, visualization, and measurement devices.

In considering what will be included in both formal coursework and informal activities, there are concepts to build and there are the materials and projects to be used with young people. The ideas listed here can be used as jumping-off points for activities.

- *Camera obscura.* Create the camera, then record an image for posterity. Interview art students to develop what's needed for pinhole camera; brainstorm design ideas; develop a mathematical model for the relationships between the size of camera, aperture and target object, and distance from the camera; build-test-refine camera design.
- *Building.* The focus is on creating workable designs for bridges, tall buildings, and other structures. This combines practical building experience with scientific principles as well as creativity. One possibility: form groups who are given a stack of paper and tape. Create a building that can still stand when a stack of books is placed on top of it.
- *Environmental issues.* Students brainstorm issues to be considered, and then research the problems. Designs are developed to address these concerns.
- *Flying machines.* Of particular interest in the age of drones, young people learn principles of aerodynamics to create simple machines.
- *Cardboard car.* Young people can be encouraged to learn about the mechanics of automobiles, then use cardboard to develop new designs for auto bodies, focusing on aerodynamics, industrial design, and aesthetics.

RESOURCES FOR ESTABLISHING ENGINEERING INITIATIVES

Most states have initiatives around engineering education in K–12 settings. They typically provide information about engineering in schools; maintain relationships with organizations and agencies that can provide guidance and expertise. Check your state's education department website to see what is available. Some examples include:

- Ohio STEM Learning Network (http://www.osln.org connects non-education) organizations with schools to facilitate adding engineering for middle and high schoolers.

- Massachusetts STEM Nexus (http://www.mass.edu/stem/home/stemplan.asp) developed a STEM plan for the state that identifies partners, coordinates funding, and works with the Department of Elementary and Secondary Education, which has developed curriculum and other resources for schools (http://www.doe.mass.edu/stem/).

- North Carolina Department of Public Instruction (http://www.dpi.state.nc.us/stem) provides a strategic plan and resources for STEM education, including rubrics for program implementation.

- The Colorado Department of Education (https://www.cde.state.co.us/stem/resources) provides links to content-rich sites for educators, parents, businesses, and students.

- The Washington (state) Office of Superintendent of Public Education lists basic definitions of STEM terms, a directory of "lighthouse" schools and links to other agencies, including Washington Laser (https://www.wastatelaser.org) and Washington STEM (http://www.washingtonstem.org).

- In Canada, check with provincial education offices for assistance in STEM program development. Isha DeCoito (2016) has done a review of STEM development in Canadian schools. Ghislain Samson (2014) has also outlined barriers in teacher education to the use of interdisciplinary efforts. Check with provincial education departments for more information on engineering in schools.

Locally, developing an engineering program begins with involving a group of educators and community members to build a plan and a program. In the role of instructional consultant the librarian can be a key participant in this process. Connect with science teachers and administrators who are exploring this possibility or open a discussion with them about including engineering in the school curriculum. If there is a working group in the district, offer to participate. Part of the rationale to present is that the library can offer support for the program.

Once an internal mandate for developing a program is in place, look to resources outside the school to help build the program. This not only makes the program stronger, but the outreach can help with community support and with budgetary issues. Involve the public library in the process (and vice versa). Connect with local businesses directly. Work with the Chamber of Commerce or other local business groups. They can offer everything from expertise in equipment needed to speakers, mentors, and financial considerations and backing.

Common Programmatic Elements for Engineering Programs

Engineering in middle and high school programs builds on a base of mathematical and scientific concepts using a systems engineering approach. It moves beyond either math or science, using creativity to create new products or processes. Schools may take advantage of prepackaged engineering curricula as a way to start programs. This is pragmatic in terms of getting the program off the ground. Care should be taken in implementing the program in a way that supports community needs, not just as a quick way to get the program going. Some factors to consider:

- Employ coaching as an instructional approach, rather than the traditional teacher role of lecture and drill. Guidance is given by modeling, encouraging, and questioning.
- Create practical experiences implementing math and science concepts to create a product.
- Experiment with a systems engineering approach, where students move from theoretical to hands-on application. They will practice the mindset of engineers, like teamwork, process protocols, and tools needed to complete the work.
- Provide experiences with practicing engineers to build a picture of what an engineering career would involve and the types of work goals they employ.

In the development of engineering coursework in schools, educators have come up with some great new approaches. Two possible—and popular—directions to go in implementing engineering in schools include the "fab lab" and using robotics as a unit of study, full course, and as a competitive extracurricular event.

Often schools opt to work with universities in developing these programs, taking advantage of the expertise found there. Some universities and science museums have implemented hands-on engineering experiences that can provide models for schools and provide presenters for events in both school and public libraries.

Fab Labs

Engineering offerings in schools most often grow within the science department. Often a retooled science lab or other classroom space supports engineering courses and/or activities. Dubbed the "fab lab," short for fabrication laboratory, this space provides activity centers where students work through the centers to solve engineering concepts, ultimately choosing a real-world issue and developing possible solutions. Fab labs have a collection of commercially available computer-controlled machines and tools, with a focus on invention and production.

Fab labs are usually equipped with industrial-grade fabrication and electronics tools. The investment for such a lab can be sizable. Planning for the fab lab may involve a multiyear development, systematically implementing the resources for the program over an extended time to spread out costs. This also helps for future replacement, as tools will not wear out or become obsolete all in one year. Outside capital resources may need to be tapped to fully furnish a fab lab, whether an individual donor or contributions from local businesses in cash or in-kind services. Equipment that can typically be found in a fab lab includes:

- Computers with design software and other software to support work in the lab. There are open-source software solutions that can be used.
- 3-D printer
- Laser cutter used to make two- and three-dimensional structures
- Sign cutter used to make flexible circuits and crafts
- High-precision milling machine to make circuits and molds for casting
- Wood router for building furniture
- Electronics for developing circuits and programming micro controls

Short of a classroom fab lab, the lab could use space in the library for independent work on projects, gathering specialized equipment in a space available to all students. In some schools, the term "innovation center" has been used, extending the term makerspace to involve innovative practices in all school disciplines. Elements like 3-D printers and robotics kits can be part of the makerspace. Having the equipment in the library would facilitate workshops and other informal approaches both during and outside the school day. Because engineering is a new endeavor for schools, having the fab lab open to the public for presentations and hands-on work can build support and understanding, and lay the groundwork for a full fab lab.

Where can information be found on establishing a fab lab? An early resource is the Massachusetts Institute of Technology's Center for Bits and Atoms, established in 2002. The FabFoundation site includes basic information on starting a lab, including conditions they deem essential for a facility to claim the name "fab lab." At the conclusion of this chapter, find program and resource ideas for engineering instruction that include fab lab resources.

Robotics

Robotics has become a popular vehicle for engineering instruction, as well as for extracurricular activities. It reaches down to elementary grades, particularly through programs like the Lego® Mindstorms® engineering program. As an extracurricular activity, there are robotics competitions for student participation at all age levels. If a school or public library already has such an informal program at the elementary level, it is a great building block to extend robotic activities and coursework to the secondary level.

Robotics involves multiple disciplines, and is often connected to mathematics programs. This field has elements of electrical and mechanical engineering, as well as computer programming, information processing, and fields like biomechanics. Given its diverse roots, it would be a perfect mode for interdisciplinary instruction involving science and math instruction. The library could be the setting for collaboration among students as projects are developed.

For robotics programs, there are a number of resources that are commonly used. Arduino and Raspberry Pi are open-source coding and robot tools that are popular in school and library settings. Other robotic kits are sold commercially and have the benefit of including all the resources a librarian needs to get started with robotics. (See the Vendors list in Appendix D.) If you are forging ahead with a robotics event, keep in mind that there may be young people among your target audience who are already proficient and can help with your learning curve as well as coaching and demonstrating the use of the robotics tools. In a school setting, check in with science and math teachers to see if they have already implemented a robotics unit in their instruction. They, too, can be mentors for you.

Cooperation with Other Educational Institutions

Within the community or region, identify outside institutions that can support the work being done in schools and libraries. This would include local museums: science museums, museums focused on local industry, and children's museums usually have activities and exhibits that are already using design work with children and young people. If the area is more rural, explore ways to connect online with facilities doing engineering work and activities.

Local vocational centers, community colleges, and universities will have experts who can advise on setting up projects for young people. Professionals in these organizations can also provide valuable input into developing curricula and into the resources needed to build a program. They may also be enlisted to be leaders in engineering activities for young people. There may be options for mentoring or shadowing activities for young people. This book includes appendices that highlights organizations that may be helpful in developing a contact list (See Appendices A and B).

RESOURCES FOR ENGINEERING ACTIVITIES

These websites feature high-quality information pertaining to engineering initiatives that can help in planning programs in school and public libraries, from single activities to full curricula. Product information can be found in Appendix D of this book for fab lab and robotics resources.

Carnegie Mellon Robotics Academy. http://education.rec.ri.cmu.edu
> The University has a wide range of resources, from curriculum to teacher training to competitions for students at all grade levels. The site includes documents and some free versions of curricula to use, with for-purchase versions also available.

Center for Bits and Atoms. http://cba.mit.edu
> Review projects, papers, and events around engineering. Some project links include descriptions of equipment that could be part of a fab lab. The Center is a Massachusetts Institute of Technology (MIT) facility. The FAQ section is a good place to start. http://fab.cba.mit.edu/about/faq

CK-12. http://www.ck12.org/teacher/
> This organization focuses on open-source, adaptable textbooks, allowing for audio and video components, along with support materials like an implementation guide. There are prepackaged texts and users are free to adapt them for their uses. Accounts are free. An overview video explains the concept: https://youtu.be/jbN-fPQnBeQ. Search "engineering" to see available materials.

Cleveland Municipal School District. MC2STEM High School. http://www.mc2 stemhighschool.org
> This school district charter focuses on project-based learning and STEM in coursework and out-of-school experiences. The website includes descriptions of projects.

Community for Advancing Discovery Research in Education. "Spotlight on K-12 Engineering Education." http://cadrek12.org/spotlight-engineering
>CADRE provides a list of resources funded by the National Science Foundation, including studies, virtual poster projects, videos about engineering experiences in schools, and project ideas.

DIY; Awesome Skills for Awesome Kids. https://diy.org
>The DIY site is designed for kids seven and older to explore projects and post their own efforts. Check the FAQs section for guidelines about the site.

Fab Foundation. http://www.fabfoundation.org
>A clearinghouse for all things fab: definitions, materials to include, and advice on how to create a fab lab. Other universities have adopted this approach: Stanford has a FabLearn program.

FabLabTV. http://www.fablabtv.com/index.html
>This program airs on Fox television (affiliate list on the site), and programs can also be found on the FabLab channel on YouTube at: https://www.youtube.com/channel/UCYWBEz-zMhkVjvNMPbXmrbg

FIRST. http://www.firstinspires.org
>FIRST (For Inspiration & Recognition of Science & Technology) sponsors competitions across North America; Tech Challenge for middle schools and Robotics Competition for high schools. The site provides information about setting up teams, volunteering, and events. For each competition area, resources and guides are available on the site.

LEGO® Engineering. http://www.legoengineering.com
>This teacher site was developed by the Tufts Center for Engineering Education and Outreach (CEEO). Find information that goes beyond the basics of using LEGO® products for engineering projects.

LinkEngineering Educator Exchange. http://linkengineering.org. This is a resource rich site designed for K–12 educators to share lessons, activities, interviews, and other resources with other educators and engineering professionals. It is sponsored by Achieve, Inc., American Society for Engineering Education (ASEE), Council of State Science Supervisors, International Technology and Engineering Educators Association (ITEA), and the National Science Teachers Association (NSTA).

Mahtomedi (MN) School District. Engineering Program. http://www.mahtomedi.k12.mn.us/page/3135
>Review Mahtomedi's K–12 engineering program and take a virtual tour of their Fab Lab: https://safeshare.tv/x/AUggSrKsAm

Massachusetts Institute of Technology. http://ocw.mit.edu/high-school/engineering/
>MIT has open courseware, some undergraduate materials, and some courses specifically designed for high schoolers on their website.

NASA. The Robotics Alliance Project. https://robotics.nasa.gov/edu/9-12.php
>NASA provides a listing of projects including both science and math connections to engineering and robotic devices.

NASA Engineering Design Process Video Series. http://www.nasa.gov/audience/foreducators/best/edp.html#.WAzzmDuueCQ

National Science Foundation. *Next Generation Science Standards HS Engineering Disciplinary Core Ideas.* 2013. http://www.nextgenscience.org/sites/default/files/HS%20ETS%20DCI%20combined%206.13.13.pdf

National Science Foundation. *Next Generation Science Standards MS Engineering Disciplinary Core Ideas.* 2013. http://www.nextgenscience.org/topic-arrangement/msengineering-design

National Science Foundation. *What Is Engineering?* Video 6:20, February 24, 2016. https://www.nsf.gov/news/mmg/mmg_disp.jsp?med_id=80126

National Science Foundation. YouTube Engineering Playlist. https://www.youtube.com/playlist?list=PL0ujJTaPsv3czgs_JoUa3QRR4cLRsY2uq

This playlist is idea central for engineering projects and viewing for young scientists and teachers.

National Youth Science Day. http://4-h.org/media

The 4-H organization hosts this annual event and provides materials for implementation.

PBSKids. Design Squad: The Design Process. http://pbskids.org/designsquad/parentseducators/workshop/process.html

This online workshop for educators and parents explains the design process. Check the PDF called "Introducing the Design Process," also available as a downloadable poster. The site was developed to accompany the Design Squad programs for middle school students.

Robotics Education & Competition Foundation (REC). http://www.roboticseducation.org

REC promotes robotics as a discipline of study and as a competitive outlet for students. The "New to Robotics?" tab is a good starting point for ideas on competitions, funding, and curricula.

Sketch Up. https://edshelf.com/tool/sketchup/ http://www.sketchup.com

Sketch Up is a 3-D modeling tool with templates to start. Education version is available as a statewide grant for K–12. For instance, in Ohio the statewide license is administered through INFOhio, the statewide network for school libraries. Contact your state educational technology office for information if it is not yet available in your school.

TeachEngineering Curriculum for K–12 Teachers. https://www.teachengineering.org

This is a free curriculum library set up through National Science Foundation funds at the University of Colorado. A free account can be set up to save choices from the offerings. It is a searchable digital library collection of standards-based engineering curricula for applied science, math, and engineering for K–12 settings. You can browse by activity category or by national standards.

Thingverse Education. http://www.thingiverse.com/education

Find projects involving 3-D printers and designing under a Creative Commons license. Free account registration enables downloading of projects on the site, and uploading of new projects. The site is an outreach of MakerBot 3-D printers.

Thinkbox. Case Western Reserve University. http://thinkbox.case.edu

The equipment list and the policies regarding use of equipment where users are required to have an orientation before using specialized equipment can be instructive for setting up labs and makerspaces.

Visioneering. https://www.smu.edu/Lyle/Institutes/CaruthInstitute/K-12Programs/Visioneering

Hosted by Southern Methodist University, this is an annual half-day event for middle school students held at the university.

REFERENCES

DeCoito, Isha. 2016. "STEM Education in Canada: A Knowledge Synthesis." *Canadian Journal of Science, Mathematics and Technology Education* 16, no 2 (April–June 2016): 114–128.

Honey, Margaret, Greg Pearson, and Heidi Schweingruber, editors. Committee on Integrated STEM Education; National Academy of Engineering; National Research Council. *STEM Integration in K-12; Status, Prospects and an Agenda for Research.* Washington, DC: National Academies Press, 2014. http://www.nap.edu/read/18612/chapter/1

Massachusetts Department of Education. Massachusetts Science and Technology/ Engineering Curriculum Framework. Massachusetts Department of Education, 2001. http://www.doe.mass.edu/frameworks/scitech/2001/0501.pdf

Massachusetts Institute of Technology. Center for Bits and Atoms. http://cba.mit.edu

Miaoulis, Ioannes. "K-12 Engineering: The Missing Core Discipline." In Senay Purzer, Johannes Strobel, and Monica E. Cardella (eds.), *Engineering in Pre-College Settings; Synthesizing Research, Policy and Practice*, 21–34. Lafayette, IN: Purdue University Press, 2014.

Moore, Tamara J., Micah S. Stohlman, Hui-Hui Wang, Kristina M. Tank, Aran W. Glancy, and Gillian H. Roehrig. "Implementation and Integration of Engineering in K-12 STEM Education." In Senay Purzer, Johannes Strobel, and Monica E. Cardella (eds.), *Engineering in Pre-College Settings; Synthesizing Research, Policy and Practice*, 35–60. Lafayette, IN: Purdue University Press, 2014.

National Research Council. *A Framework for K-12 Science Education: Practices, Cross-cutting Concepts, and Core Ideas.* Committee on a Conceptual Framework for New K–12 Science Education Standards. Board on Science Education, Division of Behavioral and Social Sciences and Education. Washington, DC: National Academies Press, 2012.

O'Toole, Kaitlyn. "Engineering Design Process." TeachEngineeering Digital Library. 2014. https://www.teachengineering.org/k12engineering/design process

Samson, Ghislain. "From Writing to Doing: The Challenges of Implementing Integration (and Interdisciplinarity) in Teaching of Mathematics, Sciences, and Technology." *Canadian Journal of Science, Mathematics & Technology* 14, no. 4 (October–December 2014): 346–358.

FURTHER READING

Afterschool Alliance. "Computing and Engineering in Afterschool." *Afterschool Alert.* Issue Brief No. 62. 2013. http://www.afterschoolalliance.org/issue_briefs/issue_comp_engineering_62.pdf

Boesdorfer, Sarah, and Scott Greenhalgh. 2014. "Make Room for Engineering; Strategies to Overcome Anxieties about Adding Engineering to Your Curriculum." *Science Teacher* 81, no. 9 (December 2014): 51–55.

Bruxvoort, Crystal, and James Jadrich. 2016. "Don't Short Circuit STEM Instruction; Exploring the Goals for Engineering and Science." *Science Teacher* 83, no. 1 (January 2016): 23–28.

Cereri, Kathy. *Robotics: Discover the Science and Technology of the Future with 25 Projects* (Build It Yourself series). White River Junction, VT: Nomad Press, 2012.

Coleman, Mary Catherine. "Design Thinking and the School Library." *Knowledge Quest* 44, no. 5 (May/June 2016): 62–68.

DeNisco, Alison. 2012. "Fab Labs: Using Technology to Make (Almost) Anything!" *District Administration* 48, no. 11 (December 2012): 34–37.

Engineering: Emphasizing the "E" in STEM Education (STEM Smart Brief). 2013. *CADRE.* http://cadrek12.org/sites/default/files/STEM%20Smart%20Engineering%20Brief%20final.pdf

Gilbert, Amy, and Katherine Wade. "An Engineer Does What Now?" *Science Teacher* 81, no. 9 (December 2014): 37–42

Gura, Mark. *Getting Started with LEGO Robotics: A Guide for K-12 Educators.* Arlington, VA: International Society for Technology in Education, 2011.

Hynes, Morgan, Merredith Portsmore, Emily Dare, et al. "Infusing Engineering Design into High School STEM Courses." National Center for Engineering and Technology Education. 2011. http://ncete.org/flash/pdfs/Infusing_Engineering_Hynes.pdf

Lacey, Gary. "Get Students Excited—3D Printing Brings Designs to Life." *Tech Directions* 70, no. 22 (September 2010): 17–19.

McCulloch, Catherine. June 20, 2016. "Elevating and Enhancing the 'E' in STEM Education." *Learning and Teaching Blog.* http://ltd.edc.org/elevating-and-enhancing-engineering-ed

Mercer, Bobby. *The Robot Book: Build & Control 20 Electric Gizmos, Moving Machines, and Hacked Toys* (Science in Motion). Chicago: Chicago Review Press, 2014.

Peterson, Tommy. "3D Printing Adds Another Dimension to the Classroom." *EdTech Magazine.* January 13, 2015. http://www.edtechmagazine.com/k12/article/2015/01/new-dimension

Song, Ting, and Kurt Becker. *Technology and Engineering Teacher* 73, no. 2 (October 2013): 30–34.

Stripling, Terri, and Beverly Simmons. "Get Students Revved Up! Robotics Bring Excitement to STEM." *Tech Directions* 75, no. 7 (March 2016): 13–17.

8

◇ ◇ ◇

ARTS: THE A IN STEAM

Creative experiences are part of the daily work of engineers, business manages and hundreds of other professionals. To succeed today and in the future, America's children will need to be inventive, resourceful, and imaginative. The best way to foster that creativity is through arts education.
    —Arne Duncan (President's Committee on the Arts and Humanities, 1)

STEM initiatives have been in place for decades, and there is governmental support of between 105 and 252 STEM education efforts at 13 to 15 federal agencies, with funding divided between K-12 and post-secondary education (Gonzalez, August 1, 2012, Summary).

New Administration/New Budget

At this writing, the outlook for continuing STEM support at the federal level is unclear. In statements to *Scientific American* as President-Elect, Mr. Trump said management of education should happen at the state and local level, instead of at the federal level (Gorman, 2016). In March, 2017, the president signed a bill called "Inspiring the Next Space Pioneers, Innovators, Researchers, and Explorers Act (INSPIRE)" to encourage girls through specific NASA initiatives (Ascione, 2017).

Adding the arts to the STEM picture is a more recent approach. STEAM embraces the power of the arts to combine across disciplines to offer students new ways of expression, and to draw in students who might not consider STEM their métier. What do we know about the pairing of arts with scientific disciplines? Robert and Michele Root-Bernstein have studied Nobel Prize winners and found a high number who had a background in arts. In collaboration with researchers at Michigan State University, the Root-Bernsteins studied groups of engineers, STEM honors graduates, and entrepreneurs, finding that involvement in arts and crafts characterize top performers in STEM fields (Root-Bernstein and Root-Bernstein, 2013).

ARTS AS A COMPLEMENT TO STEM

In the arts, a premium is placed on divergent thinking. Learners are rewarded for being outside the box, free to dream, looking at alternatives, and finding new directions. Creativity itself is not limited to the arts. Rather, the arts provide a practical outlet to express creativity. In STEM learning, most disciplines include a problem-solving process that includes considering all possible solutions. For instance, elements of the engineering design process include defining the problem and developing solutions through making and testing of models. The focus on creativity in the arts is a strong complement to solving engineering challenges. The arts can enhance the processes used to build, create, and develop products and concepts—the strong STEM skills all students need to hone. There are definite parallels between the scientific method and the habits of mind in the arts that focus on developing craft; engaging and persisting in hands-on work; and visioning, expressing and observation to communicate with others (Daugherty 2013, 10).

Expanding STEM to STEAM provides another venue for using a design or scientific process, a way to connect different disciplines in cross-curricular efforts. This makes it more likely that students will embrace and use such processes throughout their educational experience and into their careers. Both scientific and artistic endeavors employ hands-on work with a spiraling of effort in experimentation, adaptation, and reworking to reach a final conclusion, a final project.

Communicating the results of STEM efforts includes sharing the work with others. The arts offer many alternatives for such communication. Typically in art classes, there is a sharing of products, a peer reviewing process that fosters reflection on the work and the methods used to reach a satisfying artistic end. Imagine a project in engineering where students build a model: a modern dance interpretation of a scientific principle like inertia, or a musical representation of mathematical principles. Interdisciplinary activities that involve creating a product could be coupled with art and design to develop marketing plans and samples for the product launch. Using graphics or film to convey the possibilities would be a natural match to promote the work.

While scientific processes are systematic, the arts allow for leaps of imagination, sparks of creativity that could alter the direction of a project. Meshing the two can merge the imaginative impulse and the systematic process for developing products. In some art classes, the focus is on mastery of artistic conventions. While this is a part of growing arts-cognizant students, it is in the nurturing of creativity that the arts can best support STEAM efforts. Creativity can be developed and practiced using specific skills over time. The arts can emphasize skills such as:

- Approaching problems in non-systematic ways
- Using "what if" questions to look at alternative solutions
- Observing patterns, both expected and unexpected
- Combining a range of resources together to come up with novel ideas
- Employing various mediums (visual, oral, written, etc.) to communicate ideas

The challenge is in finding ways to provide cross-disciplinary experiences without sacrificing the understanding of core STEM concepts. One element that the arts can add to STEM efforts is risk taking. Artists and performers place a premium on risk taking, taking alternative approaches that might go against conventional wisdom. What a good challenge for young people to find these alternatives that can lead to new understandings.

The arts also bring a high premium on observational skills, paralleling what we see in science skills. So if in arts experiences young people are further exposed to the close examination of their subject matter as well as to the products they and their companions create, we both support STEM and bring more attention to the need for reflection and review. In the arts, whether music, artwork, or theatrical, much time is devoted to observation and precision in movement or artistic application. Transferring these skills to STEM topics can enhance the experiences of young people.

Including the arts with STEM is an issue that does create concerns for some STEM proponents. They worry that making room for yet another discipline will water down efforts to increase the impact and rigor of STEM subjects. Given school schedules and traditional approaches to education there is some merit in the argument, but it should not block STEAM efforts. Informal approaches and partnerships can ameliorate this somewhat, and begin to build prototypes for future integration of the arts in

STEM. It is important to note that while the creativity expressed in the arts may be called out as a way to add creativity to STEM, the work of those in the sciences and engineering also use ingenuity and unique, new ideas. So when art is added to STEM, look to it as a way to bring art as a component of solutions to problems in actual situations. This in turn will bring deeper learning experiences to young people (Jolly 2014).

BRAIN RESEARCH AND STEAM

As we learn more about how brain development impacts learning, educators have adopted methods to involve all learners in class work. In looking at STEAM, perhaps no researcher has had a larger impact than Dr. Howard Gardner. Dr. Gardner's Harvard research on multiple intelligences proposed that learners process information in eight different categories, and that each learner has proclivities for some over others. Gardner defined intelligence as "a computational capacity—a capacity to process a certain kind of information—that originates in human biology and human psychology" (Gardner, 6). The eight intelligences he identified are:

- Musical intelligence
- Bodily-kinesthetic intelligence (includes coordination, ability to direct movement to a goal, as in sports, dance)
- Logical-mathematical intelligence
- Linguistic intelligence
- Spatial intelligence (includes visualization, navigation, sense of position in space)
- Interpersonal intelligence (includes social skills, awareness of the actions and feelings of others)
- Intrapersonal intelligence (includes self-awareness, understanding of emotional reactions)
- Naturalistic intelligence (includes making distinctions in nature, grouping, identifying similarities and differences) (Gardner 8–19; ©2006 Howard Gardner, *Multiple Intelligences: New Horizons.* Reprinted by permission of Basic Books, a member of the Perseus Books Group)

Gardner's groundbreaking work refocused classroom instructional approaches to embrace multiple avenues to learning to better meet the needs of students. In practice, this has led to incorporation of different methods of addressing content, allowing young people to demonstrate knowledge gained in a variety of modes. By addressing multiple ways of learning, chances are greater for student engagement. This has also led to greater emphasis on evaluating students' strengths and areas for growth to tailor teaching to their needs. While Gardner did not intend his work to be used in education, he did address some variables he saw: the way in which content is presented (direct participation, modeling, oral, etc.), the agent of learning (educator, parent), and the cultural context of the learner. Each factor can determine what intelligence is favored (Gardner 2011). While Gardner did not address STEAM efforts, his framework informs practical approaches to involving students in hands-on learning to build conceptual understanding.

How might this play out in STEAM efforts? As educators work with students, it is helpful to give young people opportunities for reflection on the work they have done. Posing questions about the process used can help students focus on what techniques worked best. Having them do a think-aloud on the way in which they came to a solution or culmination of a project can include consideration of ways in which they learn best. Was it the tinkering (kinesthetic) that provided the "aha!" moment? Did they gain the most in team discussions (verbal-linguistic)? Was the use of models most helpful (visual-spatial)? Helping students identify the ways in which they learn best, as well as reviewing other options, can maximize their learning potential.

Since Gardner's work, further brain research has provided a wealth of insight into how young people learn and how educators can provide instruction that will help young people achieve. This has led to further research on learning styles, with pedagogical methods like differentiated instruction developed to better address learner needs. Acknowledging the disparate learning styles of students is a factor in making connections across curricular areas. Creating programs, lessons, and projects that incorporate brain research allows us to reach many more patrons, including those who haven't yet found a home in our libraries.

STEAM IN PRACTICE

Adding the arts to STEM content areas encourages cross-curricular collaborations with teachers, outside experts, and mentors. In both formal and informal learning environments, it gives young people broader opportunities to advance their learning and appreciation of STEAM disciplines. In preparation for developing STEAM activities, think about the programs already offered in your school and community that could inform your activities. In a school setting, talking with teachers in the arts and teachers in STEM topics separately would help identify those interested in making connections. Then bring those individuals together to begin the collaborative process of building lessons or units. Similarly in the public library, meeting with school librarians to see what may already be in place would help determine where to begin in programming.

As interdisciplinary activities are developed, it is key to consider the standards to be met in both the STEM area and in the arts area. It may be easy to enhance a science or math lesson with an artistic twist and view this as a STEAM activity. This is a good first step, doing some integration between disciplines. If connections are made between the standards for both areas, it can give more focus to the activity as a whole, deepening learning and fostering creativity. A natural place to start is with inquiry, which is included in both scientific and design processes. When young people take up that inquiry, they are involved in discovery, in a process that builds critical thinking. STEM and the arts are equal partners in the process. When both STEM and arts are being assessed as a part of the project, that creates a balance, a true STEAM effort (Riley, 2016).

This section presents two lists. The first is generic ideas for interdisciplinary connections paring arts and STEM, whether for informal learning or in library and classroom settings for collaborative teaching and learning. The second list is a sampling of project ideas feature in professional literature that can be adapted for use in both public and school libraries.

STEAM Connections

- For extended science projects, have teens create a journal, and make it visual—include sketches, drawings, particularly for observations of the process involved.
- Look for natural pairings, like geometric shapes, tessellations, and other mathematical concepts that can pair easily with art (e.g., tessellations in the work of M.C. Escher)
- For science/social issue connections use music and dance to give an emotional reaction to issues like overpopulation, natural disasters, urban sprawl, pollution, etc.
- Use an art object to bring in mathematical concepts of proportion, ratios, etc. used intuitively by artists (e.g., the golden mean or "phi" and da Vinci's *Mona Lisa* or the Japanese painting *The Great Wave off Kanagawa Golden*). This could involve partnering with an art institution as well.
- Use brainstorming to begin inquiry and develop plans for project-based learning. A graphic organizer can also be designed to identify

the main issue and supporting information, as well as notes about how a product can be developed.

- As a project, create directions for a process, from a simple project like constructing a Lego™ pyramid to a complex project like building a robot illustrating each step.
- Invite students to use music to illustrate their presentations and to correlate with concepts being studied. They can sample music from many genres, or create their own songs to illustrate what they have learned.
- Use music to have students identify and describe the main concepts for a unit of study in science or mathematics (e.g., they may wish to use rhythm to correspond to patterns in nature).
- Use robots to create artwork combining computational thinking with the creative process.
- Conduct an artist-in-residence program that features technical illustrators, sculptors, or found-art creators who use mathematical or scientific concepts in their work.
- Pair an art project using shapes created with discarded plastic bottles and their caps with a science experience about the problem of plastics in recycling. Collages can be made by painting and placing the shapes to create an image.
- Create animations for scientific processes like cell development, erosion, photosynthesis (e.g., https://youtu.be/prFaSe3s9e0, https://youtu.be/ptxCE7gL6Rk).
- Specific artists can be used to illustrate concepts in science and mathematics:
  - Mondrian and Josef Albers used geometric shapes extensively in their artwork. Combine shapes, measurement and painting, and drawing in a project. Have students use grid paper or cut shapes from construction paper or other found materials.
  - Symmetry is used by many artists and can be used to help students understand the concept.
  - M.C. Escher's work is a great jumping-off point to explore tessellations.
  - Use artworks from Picasso's cubist period and have students analyze the shapes and the angles used.
- For gaming projects, either physical board games or digital games, include specifications for art to be included in the project. Call on art teachers to help develop the checklist to be used.
- Perspective is a key element of paintings and drawings. Connect the concept of perspective in art to mathematics, and have

students identify geometric concepts, ratios, and angles (chalk art often uses great perspective).

- Building activities can involve mathematical calculations, scientific principles, and art. Set up teams to work on a challenge. For example, the artist Piet Mondrian used patterns of rectangles in many of his paintings. Architects continue to explore building design that create unusual building shapes (see https://plus.maths.org/content/perfect-buildings-maths-modern-architecture).

- Explore online sites of museums featuring art and design to make connections to activities in your school or library. Here are several examples to explore:
  - Denver Art Museum's "Building Outside the Box." http://denverartmuseum.org/edu/lesson/building-outside-box
  - MoMa (Museum of Modern Art) has a section on learning themes, including design and cubism. https://www.moma.org/learn/moma_learning/themes
  - San Francisco's de Young Museum has a section on art and science which provides full lesson plans. https://deyoung.famsf.org/calacademy/artscience

- In art and dance activities, specify geometric shapes that are to be used in creating a project, and then have students identify and describe how they incorporated the ideas in their work.

Practitioner Examples

The second list features practitioners who have reported out about successful STEAM projects that can be adapted for middle and high school–aged young people in both public and school library settings. Use them as jumping-off points to develop activities and events that fit the needs of your community or school. Not sure of your skills in STEM areas? Look for partners who can help you in these activities, whether classroom teachers, parents with particular skill sets, or local experts from organizations and institutions in your region. And don't forget to tap your students or patrons. They can also serve as great guides for others involved in the project.

Burleson, Todd. 2016. "String Theory." *School Library Journal* 62, no. 9: 34–35. (Also online at: http://www.slj.com/2016/09/programs/string-theory-projects-merging-math-art-and-making/ subtitled "Projects Merging Math, Art and Making.")
Burleson describes using geometry, graphing, parabolas, and art techniques to teach string theory. The author works with 4th graders, and the projects could easily be adapted to older students.

Dabbs, Kim. "Engagement and Impact: Design Thinking and the Arts." May 20, 2014. http://www.edutopia.org/blog/engagement-impact-design-thinking-arts-kim-dabbs

This blog post describes how the West Michigan Center for Art + Technology (WMCAT) developed an after-school program focused on creating apps, and incorporating arts into STEM processes.

Education Closet. STEAM Education. https://educationcloset.com/steam-education/

This article provides a collection of teacher postings featuring the integration of arts into STEM activities.

Exploratorium. "Arts." https://www.exploratorium.edu/explore/arts

This San Francisco–based science museum has an extensive collection of educator resources, including a section on arts connections that include lessons, videos, interviews, activities, and illustrations.

"Growing from STEM to STEAM; Tips to Team Up the Arts and Sciences in Your Classroom." The Kennedy Center ArtsEdge. https://artsedge.kennedy-center.org/educators/how-to/growing-from-stem-to-steam. This article highlights several programs in the San Diego schools, with 3rd through 5th graders, which could be adapted to middle and high school grades. Practical tips are provided for making the leap from STEM to STEAM.

Hegedus, Tess, Verónica A. Segarra, Tawannah G. Allen, Hillary Wilson, Casey Garr, and Christina Budzinski. "The Art-Science Connection." *Science Teacher* 83, no, 7: 25–31. A university team partnered with a high school biology teacher to provide eight science investigations for students. The article describes some of the investigations and the art components are included, with supporting materials.

Helfferich, Deidre, Jan Dawe, Zachary Meyers, and Nancy Tarnai. *STEAMpower: Inspiring Students, Teachers and the Public.* 2014. University of Alaska Fairbanks. Agricultural & Forestry Experiment Station. http://www.uaf.edu/files/snre/publications/misc/MP-14–13.pdf

The authors provide examples of life science experiences that incorporated the arts as models for informal programming.

Imagination Foundation. "Global Cardboard Challenge." http://cardboardchallenge.com

While this challenge usually involves younger children, the concept of using cardboard to create 3-D creations using this basic material illustrates science and math concepts. Examples of projects in a variety of settings are described in the materials on this site, whether used for the actual challenge or to develop activities in your setting.

Koester, Amy. "All Things STEAM." The Show Me Librarian (blog). http://showmelibrarian.blogspot.com/p/about-me.html.

As Youth and Family Program Supervisor at the Skokie (IL) Public Library, Koester has developed STEAM activities for young people at all ages. Each link goes to a description of the individual program listed.

Koester, Amy. "Get STEAM Rolling!" *Children & Libraries: The Journal of the Association for Library Service to Children* 12, no. 3 (2014): 22–25.

Koester details ways to integrate STEAM activities in a public library program, from one-time events to clubs to a STEAM fair.

Krigman, Eliza. "Gaining STEAM: Teaching Science through Art." *U.S. News & World Report.* February 13, 2014. http://www.usnews.com/news/stem-solutions/articles/2014/02/13/gaining-steam-teaching-science-though-art
    Nettrice Gaskins uses a technique she calls "culturally situated art-based learning" to connect science concepts with the arts. One example: students study Mimbre art and create their own designs based on mathematical concepts discovered in the art.
"LTU and DPS Launch Comprehensive Program to Encourage Students in STEAM." August 14, 2015. http://detroitk12.org/content/2015/08/14/ltu-and-dps-launch-comprehensive-program-to-encourage-students-in-steam/
    Lawrence Technological University (LTU) and Detroit Public Schools (DPS) created the "Blue Devil Scholars" program that highlights arts in STEM offerings, with collaboration between faculty members and some coursework that will provide post-secondary credit.
National Aeronautics and Space Administration. Marshall Space Flight Center. "Space School Musical." https://discovery.msfc.nasa.gov/musical/index.cfml
    This "hip-hopera" was a part of NASA's Discovery and New Frontiers Programs. The video is available on this site, and may spur creations in your school or library.
National Gallery of Art. "Lessons and Activities." http://www.nga.gov/content/ngaweb/education/teachers/lessons-activities.html
    This teacher section of the gallery has some lessons that combine art and mathematics (counting, angles) and science (ecology). In each category, several explorations of specific artists are included.
Ochterski, Joseph, and Lisa Lupacchino-Gilson. "Getting an A in STEM; Beginning a STEAM Collaboration Between Art and Chemistry Students." *Science Teacher* 83, no. 7 (2016): 39–45.
    The authors detail three projects: drawing and chromatography, molecular structures and landscape drawings, and a periodic table of molas. They focus on the consumer product conjunction of science, art, and design. They provide concrete examples of standards correlation for the lesson plus a sample rubric for one of the projects as they discuss the work they did. This can be a model for how to build at STEAM project.
Rhode Island School of Design. STEM to STEAM Case Studies. http://stemtosteam.org/case-studies/
    The Rhode Island School of Design (RISD), a longtime champion of STEAM education, gathered together a diverse sampling of STEAM projects from schools, institutions, and non-profit organizations. Topics range from climate change to photography to geometry and art.
Robelen, Erik W. "STEAM: Experts make case for adding arts to STEM." *Education Week* 31, no. 13: 8. http://www.edweek.org/ew/articles/2011/12/01/13steam_ep.h31.html
    At the Dayton Regional Stem School (DRSS), art teacher Jenny Montgomery collaborates with other teachers to create cross-disciplinary work. She has worked with a biology teacher to have students create watercolor

paintings of cell structure, and has collaborated on projects to solve engineering and science issues. An interview with Ms. Montgomery can be found at: Rollins, Ron. "Art helps students take risks; it nurtures 21st century skills." *Dayton Daily News*. February 9, 2015. http://www.mydaytondailynews.com/news/news/opinion/art-helps-students-take-risks-it-nurtures-21st-cen/ndHKW/

Waters, Patrick. *STEM to STEAM*. October 10, 2016. SmartBrief. http://smartbrief .com/original/2016/10/stem-steam. Waters describes how he turned his classroom into a collaborative space for STEAM projects. The class project to invent a new chocolate bar involved science and math concepts, as well as designing labels and marketing the project through videos. Waters also involves community partners in his projects.

Zalaznick, Matt. "Putting the "A" in STEAM." *District Administration* 51, no. 12 (2015): 62–66. https://www.districtadministration.com/article/putting-"a"-steam

The author features the Tukwila School District near Seattle, WA, where a concerted effort has been made to infuse design and technology into all disciplines. Specific examples are given about the connections: pairing a book about squirrels stealing nuts from birdfeeders with the design, building and testing of students' feeders and a high school pairing of Russian music with lessons on the science of hearing.

ADDITIONAL RESOURCES

Allan, Andy. "How Art Motivates STEM Learning." September 12, 2016. National Afterschool Association. http://naaweb.org/professional-development/item/568-full-steam-ahead?utm_source=September+15%2C+2016&utm_campaign=March14&utm

Anderson, James. "The Thoughtful Teacher." 2014. http://habitsofmind.org/wp-content/uploads/2014/08/The-Thoughtful-Teacher.pdf

As a consultant, Mr. Anderson works with Costa and Kallick's work on habits of mind. This article focuses on the importance of mindset for students.

Art Institute of Chicago. "Science Art & Technology." 2003. http://www.artic.edu/aic/education/sciarttech

Segments of a year-long course for teachers, with video clips, materials, and links to subsequent guides and lesson plans created by participants.

Bailey, Curt. "An Artist's Argument for STEAM Education." *Education Digest* 81, no.1 (September 2015): 21–23.

Daugherty, Michael K. "The Prospect of an 'A' in STEM Education." *Journal of STEM Education*, 14, no. 2 (April–June 2013): 10–14.

Five-Minute Film Festival: Arts Integration Turns STEM into STEAM. http://www.edutopia.org/blog/film-festival-stem-vs-steam

Jemison, Mae. Teach Arts and Sciences Together. *TED2002*. February 2002. https://www.ted.com/talks/mae_jemison_on_teaching_arts_and_sciences_together

Astronaut, doctor, dancer, and art collector: Mae Jemison provides a convincing talk about the intersections of science and art.

Maeda, John. How art, technology, and design inform creative leaders. June 2012. TEDGlobal 2012. http://www.ted.com/talks/john_maeda_how_art_technology_and_design_inform_creative_leaders

The former president of the Rhode Island School of Design explores the connections that make art and design important components of leadership.

Maeda, John. "STEM to STEAM: Art in K-12 Is Key to Building a Strong Economy." *Edutopia*. October 22, 2012. http://www.edutopia.org/blog/stem-to-steam-strengthens-economy-john-maeda

Palm Beach County (FL) Primetime. STEAM Initiative. STE[a+]M. http://www.primetimepbc.org/steam-initiative

Community resources are featured for after-school activities, ranging from discussions of what STEAM is to programming to a blog to resources for those who don't feel they are experts in science and mathematics.

Pink, Daniel H. *A Whole New Mind; Why Right-Brainers with Rule the Future*. New York: Riverhead Books, 2006.

Pink's eminently readable book could serve as a good professional learning group read. He posits six qualities inventors and artists share that can enrich the world, and how to develop those qualities.

Root-Bernstein, Robert S., and Michele M. Root-Bernstein. 2001. *Sparks of Genius*. 2001. New York: Mariner Books.

The Root-Bernsteins highlight 13 thinking tools based on historical figures considered to be highly creative, from Leonardo da Vinci to Jane Goodall with an eye to how educators can use these tools. This could be used for a professional learning group discussion.

STEM + Art = STEAM. https://www.youtube.com/watch?v=l3e7c1K_aDk

This video focuses on how the arts support scientific inquiry via a STEAM exhibit in Westchester, NY.

Waters, Patrick. "STEM to STEAM | SmartBrief." October 10, 2016. http://smartbrief.com/original/2016/10/stem-steam

Waters provides his view on STEAM and how it impacts his classroom, as well as a listing of resources he uses.

Wynn, Toni, and Juliette Harris. 2012. "Toward a STEM + Arts Curriculum: Creating the Teacher Team." *Art Education* 65, no. 5: 42–47.

The article discusses ways in which individual teachers have made deep connections between art concepts and science and math activities.

REFERENCES

Ascione, Laura. "New Trump Laws Will Support Women in STEM Fields." *eSchoolNews*. March 27, 2017. https://www.eschoolnews.com/2017/03/27/new-trump-laws-will-support-women-in-stem-fields/

Daugherty, Michael K. 2013. "The Prospects of an 'A' in STEM Education." *Journal of STEM Education*. 14, no. 2: 10–14.

Gardner, Howard. *Multiple Intelligences; New Horizons*. New York: Basic Books, 2006.

Gonzalez, Heather B. and Jeffrey J. Kuenzi. "Science, Technology, Engineering, and Mathematics (STEM) Education: A Primer." Congressional Research Service, August 1, 2012. https://fas.org/sgp/crs/misc/R42642.pdf

Gorman, Christine. "How President-Elect Trump Views Science." *Scientific American*. November 9, 2016. https://www.scientificamerican.com/article/how-president-elect-trump-views-science/

Jolly, Ann. "STEM vs. STEAM: Do The Arts Belong." *Education Week*. November 18, 2014. http://www.edweek.org/tm/articles/2014/11/18/ctq-jolly-stem-vs-steam.html?tkn=YWND/0gO/NlW6CUW48kknDeHqoF0t4aRuI5h

President's Committee on the Arts and Humanities. *Reinvesting in Arts Education: Winning America's Future Through Creative Schools*. Washington, DC: President's Committee on the Arts and Humanities 2011. http://www.pcah.gov/sites/default/files/photos/PCAH_Reinvesting_4web.pdf

Riley, Susan. "Is It Arts Integration or STEAM?" *Education Closet*. November 30, 2016. https://educationcloset.com/2016/11/30/arts-integration-steam/

Root-Bernstein, Robert, and Michelle Root-Bernstein. 2013. "The Art & Craft of SCIENCE." *Educational Leadership* 70, no. 5 (February 2013): 16–21.

9

◇ ◇ ◇

MAKERSPACES

PLANNING AND IMPLEMENTATION

Create! Play! Problem-solve! Make something new! It has been said that "play is the work of children" (Jean Piaget, https://www.brainyquote.com/quotes/quotes/j/jeanpiaget751100.html), but we never outgrow the excitement of creating something unique, especially if our creations are done without the pressure of grades or employment evaluations. Proponents believe that the makerspace movement changes the purpose of the library, changing passive learning into active, hands-on learning (Rendina, AASL, 2014). An added bonus is that it can often reach those young adults who learn in a non-linear fashion, those who are kinesthetic learners who learn best by "doing" rather than seeing or listening, and those who learn best by collaborating with others (Martinez, 2014, 1). This is what a makerspace can do for your library—bring back the fun of creativity to your pre-teens and teens and transform learning from teacher-centered to student-centered (Maxwell, 2015, 13). What is a makerspace? Does your library need one? If so, how will you start? How will you gain the funding needed?

Sample Needs Assessment

Makerspaces work best when the vision is tied to the unique needs of *your* own library and community (Hough, 2014, 16). No two libraries will have the same needs or the same type of makerspaces. Some makerspaces are high tech, while many others are simple and very low tech. Some occupy entire rooms, including 3-D printers, a variety of circuit boards, robotic kits, and other high-tech equipment. Others are comprised of tables in a small, dedicated area with building blocks, coloring books, craft materials, or art supplies. In this chapter, you'll learn about the many types of makerspaces as well as the resources and vendors available. (See Appendix C: Makerspace Supply List and Appendix D: Vendor List.)

If you want an overview before you start to analyze your own needs review Figure 9.1, Astrid Poot's chart called "Making Makers," which gives you ideas for starting out; it can also be used for creating badges as a sign of accomplishment of skills learned (http://lekkersamenk looien.nl/wp-content/uploads/2017/02/makersmaken_50tools_ENG .pdf). (For more on badging see Chapter 11.) A helpful checklist the "100+ Makerspace Products & Materials" is free but requires that you register your e-mail address (https://www.makerspaces.com/ makerspace-materials-supply-list/).

Analyzing Your Library's Needs

First, assess the needs of your library. Find a group of staff members who are willing to brainstorm with you. Start with your department or create your own group of "consultants" choosing only those with a positive attitude, a sense of humor, and the ability to be honest with you. This core group becomes your sounding board. Initially you want it to be made up of the adults you work with; we'll get to our young adults in a minute. These are the players who will help you to identify potential obstacles and solutions, to evaluate the process, and to determine what needs to be changed. Your staff can be defined as your department in a public library, your assistant, and a core group of teachers if you are in a school library. Once you have some answers to the questions posed in Figure 9.2 and have identified your patrons' needs consider adding your tech and maintenance staff members to your group as they will have unique perspectives to add.

Figure 9.1: "Making Makers" chart
Courtesy of Astrid Poot.

From *Full STEAM Ahead: Science, Technology, Engineering, Art, and Mathematics in Library Programs and Collections* by Cherie P. Pandora and Kathy Fredrick. Santa Barbara, CA: Libraries Unlimited. Copyright © 2017.

Questions to discuss with your consultants at the planning stage:

1. What are the needs of our patrons? (You will do a survey of patrons later.)
2. What do we want them to be able to do? (Jakes, 2016, 1)
3. What does the library already have that can be re-purposed? (Check other departments.)
4. What do patrons request that we can't yet provide?
5. What national, state, or school standards can be met?
6. Do we want a low tech or a high tech space?
7. What do you need to create it and what will it cost?
8. What experts do we have on staff in the library? In the school?
9. How will we schedule the space or cart?
10. Will we use a drop-in model on students' free time (informal learning) or will it be used for group/class projects (formal learning) as well? (Jakes, 2016, 2)
11. Will we start with after-school programs, such as Jensen's Maker Mondays?
12. What ground rules will we set?
13. What space is available?
14. How will we fund your makerspace?
15. What are the resources we need? See Sample Makerspace Supply List in Appendix C.
16. What vendors will we use? See Vendor List in Appendix D.
17. Are there any Makerfaires or Mini-Makerfaires scheduled nearby?
18. How will we measure success—anecdotal and statistics, training, staffing, outreach. (See Chapter 3: Outreach and Collaboration; Chapter 11: Evaluation and Measurement)

Figure 9.2: Consultant Questions

Analyzing Your Patrons' Needs

Since you are designing the space for teens and tweens you need to gather information from them. This is Step 2. Their requests and insight may surprise you. If you have a Teen Advisory Board meet with them and record their suggestions before creating the survey. They serve as the focus group to review questions for your teen survey. Follow their suggestions as to wording and delivery systems, that is, getting the surveys into the hands of their friends.

Create a short survey (online, in person, or on paper) of no more than five questions. Feel free to adapt questions from those used with your consultant team. Make all surveys anonymous with open-ended spaces for their opinions and suggestions for activities or speakers. Lead them to the survey by postings on social media, announcements in the school, and placing them on the tables. In the survey for young adults, eliminate questions that deal with the mission, budget, and philosophy. Be sure to ask if any of them are experts in a particular skill as you may find numerous experts among your patrons, their parents, or their neighbors and friends.

Pick a day to observe your patrons and how they use the library. Have other staff members in your department do so on a different day. Warning: be careful that you don't look like you are stalking them or waiting to "catch them" doing something wrong. Just do a simple "walkabout" in a meandering sort of way. Do not stop unless they have questions but pay attention to what they are doing. Are they reading? If so, are they into dystopian novels, fantasies, zombies? Are any of them relaxing—coloring, knitting, sketching buildings, playing computer games on their phones, listening to music? If you do chat with them, record your notes *later*. (They may become nervous if you are writing while they are talking, that's too much like a classroom evaluation.)

Compare your observation notes with your staffers and then consider these questions:

- Did you observe any hidden talents or skills in your walkabout?
- Did students mention something that is missing from your programming schedule?
- Did you get any suggestions from your "regulars"?
- What about your "outliers," did they offer any suggestions?
- Did you receive any suggestions from speakers or experts?
- Do any of them have a talent that they would like to share?

Once you have some answers you can draft an Action Plan. (See Figure 11.1 for a sample Action Plan.) This guideline includes goals, needs, budget, potential experts on staff and in the community, and a plan for how you will staff the area and provide security. Ask also how do we provide a variety of activities to accommodate multiple intelligences and various ability levels. How do we ensure that we have activities that will appeal to both guys and gals?

Makerfaires

If you want some ideas before designing your own spaces visit local libraries, public or school, that have their own makerspace. There are numerous makerfaires and mini-makerfaires held throughout the country on an annual basis. Barnes and Noble bookstores usually hold one every November wherein robotics, interactive pens, and circuit boards are available for customers to "play and experiment" with these tools with resource people nearby to demonstrate and assist with questions. Makerfaires are held throughout North America, Europe, and Asia and in Kuwait. Large makerfaires are held annually in Orlando, Florida and Chicago, Illinois.

San Mateo, California (the Bay area) holds a mini-makerfaire sponsored with *Make* magazine. Some locations such as Vancouver, British Columbia and Calgary, Alberta have held more than six makerfaires. Use the map on the "Maker Faire" (http://makerfaire.com/) web page to locate makerfaires by name, date, city, state/province, or country.

CREATING YOUR MAKERSPACE

Once you have formulated your action plan regroup with your consultants. Review the surveys and determine a place to start. Determine how your project-based learning (PBL) activities will enhance learning (Martinez, 2014, 2). Ask your group the following questions and promote "out-of-the–box" thinking. Designate a recorder so that all ideas, even the wild ones, are considered.

How Do You Start?

What is the one activity to tackle first? What should be next? If at all possible, come up with a three- to five-year plan that you can show to your supervisor to illustrate that you have thought out the process for longevity/sustainability.

Facility

How much space can you devote to a makerspace? Can furniture be rearranged to provide an area? Is there a little used area adjacent to the library that can be utilized? Is there unused furniture in the building that can be repurposed for your area?

Training

Having a handy set of directions as a starting point can be a help for beginning students, teachers, and new staff members or volunteers. Use directions as training tools then let the teens create and illustrate their own instruction sheets in areas of their expertise. In 2017, student Sylvia Todd, age 15, created a number of pdf files that she has shared; many include Creative Commons permission. See sylviashow.com/printables (Martinez, 2014, 2). I particularly like the one-page sheet called "How to Solder" that contains what to do and what NOT to do. Another instruction sheet she created is called "Use Soft Circuits to Electrify Your Wardrobe." Instructables.com is another place to find instruction sheets so that you don't have to create them all on your own. There will be times when students will get stuck and not know how to proceed—have handy a copy

Figure 9.3: Unlock the Maker Button

www.lekkersamenkloojen.nl
(c) Astrid Poot, 2017

From *Full STEAM Ahead: Science, Technology, Engineering, Art, and Mathematics in Library Programs and Collections* by Cherie P. Pandora and Kathy Fredrick. Santa Barbara, CA: Libraries Unlimited. Copyright © 2017.

of Figure 9.3, the "Unlock the Maker" button. Astrid Poot has created a two-sided button that provides hints for a teen who isn't sure what the next step should be.

Consumables

What supplies will you need? Are there any supplies on hand in your department? Do other departments have items they no longer use that can be utilized in your makerspace? An old storage closet may have useful, but forgotten items. Ask before borrowing. Check: Is there room in the budget to purchase supplies?

Brainstorm outside Partners

List the name of potential experts supplied by your consultants or your patrons. Note that other libraries, schools, social agencies, community organizations, businesses, and local universities may be able to provide you with speakers or participants although some may request a stipend.

Funding

What sources can you tap for additional funding? There may be local resources such as Friends of the Library groups, Trustees, PTA/PTO, Educational Foundations, and service groups (i.e., Rotary, Eagles, Kiwanis, Garden Clubs). (See Figure 2.2, Grant Sources.) In Canada check bank websites as they offer a number of local grants. These groups sometimes provide local grants and are discussed in greater length in Chapter 2.

Name and Logo Contest

Allow patrons and staff members to submit names for the new makerspace area. Have your consultant group or Teen Advisory Board whittle the selections to a few and then ask your young adults to vote. Once a name is selected add a second contest, limited to tweens and teens, to create a logo for the makerspace.

Buying Plan

Provide your consultants with the article "100+ Makerspace Products & Materials." The article reviews many items that are commonly found in makerspace labs and appends a list of consumable materials at the end, including everything from straws, fabric, and cardboard to fishing line and balsa wood, glue, duct tape, magnets, and PVC pipes (100+ supply list, 9–11). You need *not* duplicate this list, remember, it is the needs of your patrons and your library that will drive your space. It is helpful to see the list and inventory what you already have on hand. To get the free supply list you will need to provide an e-mail address and sign in to follow them on Twitter, however, the listing is very thorough, as of its writing, providing an item name, a few sentences of description, and the website for further research.

The best makerspace items are those that can be used, taken apart, and then configured differently, allowing for a variety of designs and solutions. Include items that require your patrons to work together but initially require little adult supervision. Save the soldering irons for later! Give students a design challenge (Landgraf, 2015, 32). A summer workshop once required students to create a model of a belltower using only cardboard, chipboard, paper clips, and small metal bells. Students learn from, and enjoy, those challenges that allow them the freedom to create, collaborate, and design without following step-by-step directions on an instruction sheet (Jakes, 2016, 2). Items should be able to be used by different age and ability levels. Some students will be happy to have crafting supplies while others will want to create games, animations, or movies. Still others will look forward to projects that involve circuits and robotics (Schley, 2016).

Administrative Support

Once you've completed your survey of teens and your focus group with your consultants, create your action plan. At that time you need to speak with your principal or department head, or director to gain his or her support. Determine your selling point and prepare your "elevator speech" presentation (short but to the point) before approaching your manager (Borman, 2016, 20). Be prepared and be positive and always, always, always, put forth the ways in which your makerspace will benefit your patrons (e.g., skills and knowledge they will gain). Managers need to know what you are working on and, more importantly, how it can help the library to meet its goals and make a positive impression on the community.

Your project-based learning activities provide teens with hands-on skills and a chance to think critically while collaborating with others so be sure that your talking points stress the standards or developmental standards that you will be met. You need to keep your manager informed even if you do not need money from his or her budget. They do NOT want to be blindsided. Refer back to your action plan for specifics. See also Chapter 2, which deals with presenting your case to a reluctant administrator.

If you are in a school, be sure to emphasize that you will collaborate with teachers. Identify the curricular goals that are met using maker-space materials. The International Society for Technology in Education (itse.org) provides a list of standards that can be met by a makerspace. (For a fuller discussion see Chapter 10.) Tinkering, making art, and creating are usually messy endeavors so it is imperative that your administration and maintenance staff are aware of this and embrace the clutter. You don't want a well-meaning custodian to throw out a project that is in progress.

Allowing staff to "play" with some of the items at a staff meeting will gain you some converts. Blogger Diana Rendina suggests that you list the standards you are meeting on a board so that teachers, parents, and administrators know that you are meeting guidelines in addition to allowing students to learn by creating (Rendina, AASL, 2014). In a public library you could simply supply your director with those of the developmental assets that you are meeting. Rendina works at a middle school in Florida and has created both a STEAM club that meets during the day, using a learning stations approach, and an after-school Makers Club that does more in-depth projects and design challenges. Best of all is her signage near the area reminding students and teachers to "Think Outside the Box." She firmly believes that as librarians we can foster a "maker culture" (Rendina, AASL, 2014).

Recent additions to the makerspace movement include the addition of compassionate projects, namely, those projects intended for charities, shelters, and other groups in need of assistance. Gina Seymour has instituted projects for animal shelters, for the homeless, and for hospital pediatric wards. Her students have made "toys for the local animal shelter, sleep mats for the homeless, and postcards for hospitalized children" (Seymour, STEAM + C, 2017). What an incredible way to add meaning to your makerspace and to give back to your community! It also models compassion for teens in a manner that is not at all preachy.

Donations

You will hear that no project is worth doing unless you can first get the funding. That's not necessarily the case. Finding donations can often allow you to start a project and use it to convince your managers that the makerspace will help students to learn, to problem-solve, and to meet standards.

Once you decide on the first items you need for your space we give you permission to beg, borrow, and seek donations. Many libraries start first by explaining the need at a staff meeting to gain support and ownership from colleagues. Gather some allies and work on a listing of the things you would like to gather first.

LOGISTICS

Before putting out a call for donations be sure that your library has a donation policy in place! (See Chapter 2.) You need to be able to donate elsewhere or trash items that do not meet your selection policy. Then identify the types of donations you are seeking. Do you need yarn and knitting needles, colored pencils and markers, building blocks, or Legos? Online newsletters for parents and the community are a great way to start but be sure to include information about donation times and locations, for example, donations can be taken to the front desk anytime the library is open. (Schools may designate another spot such as the main office from 8 A.M. to 4 P.M. so that items are not left in the hallway after school hours.) Be sure that this request is also posted on your web page. If you want to start with building blocks, then ask for these first; chances are there are many parents whose children have outgrown these toys and they would love to have a worthwhile place to send these items. A sample Makerspace Supply List is included in Appendix C as a starting point. You will not need every item on the list but it allows you to decide where you want to start.

Start with One Idea

Activities can be as simple as a coloring area or a table or wall of building blocks. Start small, suggests Jensen (Small Tech, 2016, 1). Some school libraries enlarge non-copyrighted coloring pages to cover an entire table allowing students to relax and color during their study hall or lunch times. Others have set up clubs for crafting cards, making books, knitting, or crocheting. Other spaces include creating games, web design, photography, rocketry, robotics, building circuit boards, coding, animation, movies, or recording songs (Maxwell, 2015, 14).

Do not be afraid to have students attempt easier items created or listed for younger students. Noel MacNeal's books use cardboard boxes to create items and help you to create 30 different kinds of puppets. Teens and pre-teens are ready for easy and relaxing tasks after a tough day of classes, exams, and grade stress. Don't we often have patrons ask us for a book they fondly remember from middle school? Do you get requests for a beloved picture book from their favorite story time? Utilizing such low-tech projects at the beginning can build confidence, particularly when you want to build skilled tasks such as knitting, crocheting, building circuit boards, or robots. Teens may enjoy making things that can be used elsewhere, for example, masks, puppets, or a puppet theater for children's story hour or for use in an elementary school.

Space and Staffing

What area can you reconfigure to create a makerspace? Some libraries have started with a Lego wall as they had no floor space to devote to such an undertaking. This consists of a mounted board on which teens can rearrange Legos. Others have made the wall a part of the circulation desk. Do you have an area that could be converted now? Is there a storage closet adjacent to the library that could provide a makerspace room or provide storage for materials? Do you want your makerspace to be mobile and travel to classes? If so, then you need to purchase a cart that can carry your materials to various areas of the building and secure them overnight.

How will you staff the area? Will patrons be allowed to use the area if there is no staff member present? Do you plan to train a core group of teens to assist and mentor other students? If carts travel to another department or classroom must a staff member travel with the cart? Some of these decisions will be determined by the level of difficulty of your items. If you are providing crafting tools you may be less concerned about safety and staffing. If, on the other hand, your cart includes a soldering gun and circuit boards, be sure that the tools are supervised by someone with experience in their use. Consider also what terms you want to use for your trainees and helpers—others have used terms such as resident tinkerers or artists-in-residence (Landgraf, 2015).

Circulating Items

One of your many decisions will be whether or not you will allow items to circulate. Can items travel to other departments, classrooms, teachers, patrons, or community organizations? If so, develop a written policy explaining your guidelines and rules including compensation for missing or damaged items. In her *Teen Makerspace Manual* Karen Jensen includes guidelines for selection, weeding, inventory, and storing materials. Included in her blog "What's in Your Teen Makerspace Manual?" are links to forms she has created for requesting supplies, training, goals, and daily reporting. Creating your own manual allows you to include forms that you've adapted for planning lessons or programs, to record rules for use, and for circulating items. Include specific rules that your library is required to follow regarding the loaning of library equipment. One suggestion is that students wishing to sign out makerspace materials must have on file a School Library Acceptable Use form signed by a parent or guardian; public libraries will need to decide if having a library card in their own name is sufficient (Jakubowicz, Makerspaces without, 2014). If you plan to circulate items you need to catalog your kits for ease in signing them out to students and/or teachers. The blog article "Makerspaces without a Space: Circulating Maker Kits for the School Library" by Collette Jakubowicz (listed as "Mrs. J in the Library" on her blog) provides suggestions. She staffs numerous buildings in her district and has created some MARC record templates that she uses to check out materials to students.

Security

Whether your Markerspace is a permanent place in your library or a mobile cart that travels you will need to balance accessibility with security. How do you allow patrons to easily access items while keeping items from "walking away"? (Rendina, Why, 2016) Discuss security issues with your maintenance team. Will you allow items to be left "in progress" or will you need to have items stored overnight? As the area has the potential to be messy during the day, you won't want to risk having items discarded accidentally by the team that cleans the school. Provide these staff members with pictures of the areas and walk through the area so they will know that a stray block or Stikbot (stick figures that are used in animation) found on the floor is not trash and is important to keep.

SHARE YOUR RESULTS WITH PRIDE

Take pictures and videos of projects in progress or completed items. You'll need to follow your library's policy about securing permission from parents if your students are under 18; however, posting pictures of active learning that is, arms only, no faces, is a boost for your makerspace and the creative energy found in your library. Use whatever social media platforms are already in use for public relations. Possibilities include LibGuides, Instagram, YouTube, blogs, Flickr, Twitter (#makered, #makerspace, #make), Tackk, school and library web, and Facebook pages (Maxwell, 2015, 13). Don't forget to send promos and pictures to your supervisors through your intranet then bask in the glory that is yours for bringing active learning to your patrons.

RESOURCES

Instructables.com. How to guides: make cake, table, folding workbench, robots, sew, woodwork. http://instructables.com

MakerEd.Org nonprofit resources

Makerfaire and Mini-makerfaire information, maps, and events. http://makerfaire.comhttp://makerfaire.com/mini

"Makerspace Materials" appended from "100+ Makerspace Products & Materials." Makerspaces.com (Retrieved December 22, 2016) pp. 1–11. https://www.makerspaces.com/makerspace-materials-supply-list/

Makerspaces.com has a shop, but also has a free eBook and project ideas (some of which are free), they're on Twitter and Facebook, and you can sign up to get e-mails from them

Rendina, Diana. Renovated Learning. Blog. http://renovatedlearning.com/2016/12/19/budget-friendly-projects-makerspace/

Todd, Sylvia. Printable instruction sheets for makers. http://sylviashow.com/printables

Twitter: *Make* magazine @make; SMCOE STEM Center @stemcentric

REFERENCES

Borman, Laurie D. "Makerspaces, Digital Literacy, Advocacy at AASL15." *American Libraries*. 47, no. 1/2 (January/February 2016): 20.

Hough, Lauren. "Makerspaces in the Library." *Public Libraries*. 56, no. 2 (March/April 2014).

ISTE Connects. "What Should Be in Your Makerspace Toolbox?" March 8, 2016.

Jakes, David. "5 Considerations for Designing Makerspaces." September 16, 2016. Blog.

Jakubowicz, Collette. "Makerspaces Without a Space: Circulating Maker Kits for the School Library." Posted in *How to Be Brave, Makerspace?* December 21, 2014.

Jensen, Karen. "Small Tech, Big Impact: Designing My Maker Space." February 1, 2016.

Jensen, Karen. "What's in Your Teen MakerSpace Manual? Forms Edition." SLJ: Teen Librarian Toolbox Blog, January 4, 2017. (MakerSpace Activity Planning Checklist and Program Planning Worksheet)

Landgraf, Greg. Making Room for Informal Learning. *American Libraries*. 46, no. 3–4 (March/April 2015).

"Makerspace Materials" appended from "100+ Makerspace Products & Materials." Makerspaces.com. Retrieved December 22, 2016, pp. 1–11. https://www.makerspaces.com/makerspace-materials-supply-list/

Martinez, Sylvia, and Gary Stager. "The Maker Movement: A Learning Revolution." July 21, 2014. ISTE. https://www.iste.org/explore/ArticleDetail?articleid=106

Maxwell, Angela. "The Maker Mindset: Curate, Create, Collaborate." *Ohio Media Spectrum*, 67, no. 1 (Fall 2015).

NMC. COSN. "NMC/CoSN Horizon Report 2016 K-12 Edition." http://cdn.nmc.org/media/2016-nmc-cosn-horizon-report-k12-EN.pdf

"100+ Makerspace Products & Materials." Makerspaces.com:1–11. Retri eved December 22, 2016. https://www.makerspaces.com/makerspace-materials-supply-list/

Poot, Astrid. "Making Makers." 2016. http://lekkersamenklooien.nl/wp-content/uploads/2017/02/makersmaken_50tools_ENG.pdf

Poot, Astrid. "Unlock the Maker" Button. 2017. http://lekkersamenklooien.nl/

Rendina, Diana. "4 Super Easy Budget Friendly Projects for Your Makerspace." December 19, 2016. Blog. http://renovatedlearning.com/2016/12/19/budget-friendly-projects-makerspace/

Rendina, Diana. Presentation at AASL: *Makerspaces and Libraries: How to Bring Some STEAM into Your Program*. Columbus, OH, 2015.

Rendina, Diana. "Why a Makerspace Is Not a Magic Cure-All for Your Problems." AASL. *Knowledge Quest* Blog, December 29, 2016.

Schley, Courtney. "Learning & STEM Toys We Love: Everything from Games to Maker Kits, Robots, and Crafts." The Wirecutter in Gadgetry, December 2, 2016.

Seymour, Gina. "STEAM + C: Adding Compassion to the Makerspace 'Because Nice Matters'." Blog. January 25, 2017. https://ginaseymour.com/

FURTHER READING

Booth, Heather, and Karen Jensen, eds. *The Whole Library Handbook: Teen Services*. Chicago: ALA, 2014.

Cunningham, Christine M., and Melissa Higgins. "Engineering FOR Everyone." *Educational Leadership* 72, no. 4 (2014): 42–47.

Dougherty, Dale. "Make Education: Remembering Seymour Papert, Tool Guides for Kids, and More." February 16, 2017. http://makezine.com/2017/02/16/education/.

Flowers, Helen F. *Public Relations for School Library Media Programs; 500 Ways to Influence People and Win Friends for Your School Library Media Center*. New York: Neal-Schuman, 1998.

The GCAA Makerspace. "Makerspace FAQ." January 13, 2017. https://gcaamaker space.wordpress.com/

Goerner, Phil. "Creating a School Library Makerspace: The Beginning of a Journey." *School Library Journal*. Online. January 19, 2015. http://www.slj.com/2015/01/technology/creating-a-school-library-maker-space-the-beginning-of-a-journey-tech-tidbits/

Graves, Colleen, and Graves, Aaron. *The Big Book of Makerspace Projects*. New York: McGraw-Hill, 2016.

Jarrett, Kevin. "Makerspace Middle School Journey: Shop Class Rebooted . . . Digitally." *Edutopia*. Blog. August 5, 2015. https://www.edutopia.org/blog/making-makerspace-shop-class-rebooted-kevin-jarrett

Loertscher, David. V. "The Virtual Makerspace: A New Possibility." *Teacher Librarian*. October 2015, 43:1.

MacNeal, Noel. *10-Minute Puppets*. New York: Workman Publishing, 2010.

MacNeal, Noel. *Box! Castles, Kitchens, and Other Cardboard Creations for Kids*. Guilford, CT: Lyons Press, 2013.

Porter, Marcia. Books, Bytes, Blog. "Lego Desk." January 7, 2017. Common Ground 2016 Presentation.

10

❖ ❖ ❖

CONNECTING STANDARDS

There are a myriad of ways to identify what characteristics and skills young people require to be successful in school and in life. In 2000, Arthur Costa's 16 habits of mind focused attention on those qualities successful people exhibited. The list includes specific learning skills like listening, questioning, and striving for accuracy. In addition, it outlined attitudinal qualities like persistence, flexibility, and humor (Costa, 2009). These habits have influenced standards documents, and the way educators work with students. Educators looked to the list to determine a focus for working with young people.

Standards are the conceptual statements educators use to guide their instruction and their interactions with students. Standards are markers for educators to assess young people in their growth and development as students and as citizens. To thrive as a nation, we need common goals for education that guide the work done in schools, in preparation for the workforce. This chapter considers

- What skill sets are needed to thrive
- Commonalities among standards
- Key elements of standards documents
- How to use the standards with administrators and colleagues to promote library work

SKILL SETS NEEDED TO
MEASURE SUCCESS

All standards areas involve a process of inquiry, dovetailing nicely with the efforts of librarians in providing learning opportunities for young people. Within the disciplines, there is a core of quantitative knowledge: mathematical facts, scientific principles, art concepts, technological literacy, and information seeking and use. All these skill sets from disparate standards are interrelated.

In science, foundational elements from mathematics, like arithmetic, algebra, and statistics are building blocks to more abstract thinking. The engineering design process requires knowledge of scientific and mathematical principles to develop a product or process. The arts bring creativity, using a variety of disciplines to develop artistic products. Information literacy grounds all disciplines in analysis and synthesis of information to solve problems and create solutions. In library instruction, standards revolve around concepts of informed citizenry, ethical use of information, and a research model. The research model is the core that aligns with other disciplines. Young people need to be able to communicate, collaborate, and evaluate their work.

In learning the conceptual, young people develop a basic knowledge base from which to draw as they put concepts into practical action. While standards include content knowledge, they go far beyond these basics. Within the standards set by professional organizations, there are core beliefs about an informed citizen, a lifelong learner, and a communicator who builds solutions based on evidence and practice.

Public Libraries: 40 Developmental
Assets for Adolescents

In considering the skill sets needed by our students, both hard and soft skills belong in the discussion. Hard skills would fulfill the content specifications within educational standards. Soft skills would apply to the processes and interactions that students have in learning experiences. To address both the intellectual and developmental needs of young people, we turn to a core document used in programming for young adult programs, *40 Developmental Assets for Adolescents*. Adolescence is a time of upheaval for young people, when physical and emotional changes collide in their growth toward adulthood. This checklist document was developed to provide a guide for healthy adolescent development. It informs a variety of disciplines in work with young people.

40 Developmental Assets for Adolescents was developed through a survey process conducted by the Search Institute of Minneapolis, Minnesota, beginning in 1993. The survey included over a million young people in grades 6–12 from more than 600 communities. The 40 assets are divided into external and internal. External assets are provided by outside forces— parents, teachers, community, and the overall environment. Internal assets include commitment to learning, positive values, social competencies, and a positive identity. The intent of the Search Institute in developing the document was to provide a baseline for building a program to promote positive behaviors in young people and in the community at large (Search Institute, 2007). It provides a framework for community coalitions to unite in efforts around issues from after-school programming to drug prevention to social media awareness. As we examine educational standards, keep in mind these developmental needs of young people in lessons, activities, and projects.

In addition to this document focused on skills for adolescents, the Young Adult Library Services Association (YALSA) has a document about the professional qualities needed for those who provide service to teens. It mirrors many of the attributes to be developed in young people, such as using current knowledge to building new understanding, respecting the ideas of others, and approaching new projects with an innovative mindset (Professional Values Task Force, 2015, 3–5). What better way to help students than to model such behaviors?

Commonalities among Standards Documents

Educational standards have commonalities throughout and between disciplines, even though vastly different groups designed them. Earlier versions of standards documents limited much of their focus to subject content knowledge. This also meant that the focus was largely on what students should know. That focus is still there in part, but is balanced by what students should be able to do. The "know and be able to do" approach is particularly notable in science and mathematics documents. The Common Core standards as a national initiative highlighted this refocus.

Standards now marry content with a focus on process, using terms such as: design process, scientific process, and research process. Creativity and systems thinking are key to building understanding. There is also an acknowledgement of the need for individual engagement in learning: the ability to collaborate as part of a working group, and the ability to communicate effectively within the group and with those outside the group in public settings, both physical and virtual.

STANDARDS TO CONSIDER FOR STEAM

Why this shift in approach? It grew out of continuing research on learning and teaching, which has shown that foundational knowledge—the facts and figures of a discipline—are only a starting point to deep learning. By continually building and revising understanding as young people develop, we create deeper and more meaningful learning experiences over time. Research on the learning process has shown that the more senses that are used in educational activities, the more likely it is that all students can find engagement. For the purposes of this overview of STEAM support for libraries, the key standards of the following groups will be discussed:

- National Science Teachers Association (NSTA), informed by the work of the National Science Foundation
- International Technology and Engineering Education Association (ITEEA)
- International Society for Technology in Education (ISTE)
- National Council of Teachers of Mathematics (NCTM)
- American Association of School Librarians (AASL)
- National Coalition for Arts Standards (NCAS)
- Council of Chief State School Officers (CCSSO) under the auspices of the National Governors Association Center for Best Practices

This chart (Figure 10.1) provides a quick overview of the discipline, source, process, and model/method used for educational standards.

What are the basics of the major standards that guide STEAM efforts? Most have been updated or are in the process of revision at this writing. Check with subject teachers or curriculum directors for copies of the documents, or add standards documents now available in the library professional, education collection, and on the library website. Some are readily accessible online; others are only available through purchase.

- Common Core (2010) including Mathematics and Scientific and Technology Literacy—available online
- Technology Standards for Students (2016) International Society for Technology in Education (ISTE) —available for purchase
- Next Generation Science Standards—National Science Teacher Association (NSTA)—available for purchase
- Principles and Standards for School Mathematics (2000)— National Council of Teachers of Mathematics (NCTM)—available for purchase

- Standards for the 21st-Century Learner—American Association of School Librarians (AASL)—available online
- Standards for Technological Literacy—International Technology and Engineering Education Association (ITEEA)—available for purchase
- National Core Arts Standards—National Coalition for Arts Standards (NCAS)—available online

Library standards represented in the Standards for the 21st Century Learner are interwoven in the discussions of each STEAM area standard. This is particularly true of the research process, but includes use of information, synthesis of new learning, and communicating that learning through creative projects. Within the standards representing STEAM areas, the connections to library standards will be highlighted. Knowledge

Standards Documents Overview			
Discipline	Source	Process	Model & Method
Science	NSTA	Scientific inquiry	Scientific models built on knowledge of scientific concepts
Technology	ITEEA	Engineering design	Technological models using tools to construct knowledge
Technology	ISTE	Design process	Design models using computational thinking to find solutions
Mathematics	NCTM	Problem solving	Mathematical models using analysis and mathematical concepts in solutions
Library	AASL	Information seeking	Research models using analysis and synthesis of information collected to create new meaning
Arts	NCAS	Artistic expression	Artistic process using creativity for enhance creation
Common Core Mathematics	CCSSO	Problem solving	Mathematical models using analysis and mathematical concepts in solutions
Common Core Scientific & Technological Literacy	CCSSO	Design experimental process	Design models using scientific and technological concepts to create and communicate products

Figure 10.1: Standards Documents Overview

Most states have adopted STEM standards, often within science and math standards for K-12 education. Make sure to check to see what is available for your state or region. Some examples are listed below.

Massachusetts was an early adopter of STEM standards, starting around 2001. Their website has comprehensive information about programs in the state that may be helpful in other areas, from model curricula to assessment to professional development opportunities. The document "An Effective Standards-Based K-12 Science and Technology/Engineering Classroom" outlines the qualities present in a strong program, from student and teacher perspectives. This 2008 document was followed by the 2016 document "Science and Technology/Engineering Education for All Students: A Vision" showing the present approach to STEM in Massachusetts classrooms. In addition, the Massachusetts Department of Higher Education has developed the site STEM Nexus with resources for professional development, model programs and resources.

North Carolina's approach was to collect all information in one portal. It developed the NC STEM Center as a part of the NC STEM Learning Network, a collaborative effort between the NC Science, Mathematics & Technology Education Center; the NC Department of Public Instruction; the NC Community College System; and the UNC General Administration. It houses standards information, exemplary schools, and electronic newsletter (subscription via email), and learning opportunities for North Carolina students both in-school and outside of school.

A resource that highlights STEM efforts across a number of states across the country is the Washington STEM. The advocacy organization provides a good jumping off point to investigate STEM and STEM resources in a specific state or region.

REFERENCES

Massachusetts Department of Elementary and Secondary Education. "An Effective Standards-Based K-12 Science and Technology/Engineering Classroom." 2008. http://www.doe.mass.edu/STEM/Standards-BasedClassroom.pdf#search=%22characteristics%22.

Massachusetts Department of Elementary and Secondary Education. "Science and Technology/Engineering Education for All Students: A Vision." 2016. http://www.doe.mass.edu/frameworks/scitech/2016-04/Vision.pdf

Massachusetts Department of Elementary and Secondary Education. Science, Technology/Engineering and Mathematics (STEM). http://www.doe.mass.edu/stem/

Massachusetts Department of Higher Education. STEM Nexus. http://www.mass.edu/stem/home.asp

Massachusetts Science and Technology/Engineering Curriculum Framework. Massachusetts Department of Education, 2001. http://www.doe.mass.edu/frameworks/scitech/2001/0501.pdf

NC Stem Center. https://www.ncstemcenter.org

Washington STEM. National STEM Movement. http://www.washingtonstem.org/Our-Impact/National#.WH54bXfMyb8

Figure 10.2: Standards: A Sampling of State Resources

of standards in various disciplines can be used to begin collaborative work with classroom teachers to provide thoughtful and rigorous learning experiences for young people. Each STEAM area standard document will be discussed in the remainder of this chapter. Figure 10.2 provides some background information on how some states have developed their STEM standards based on the national standards documents discussed in this chapter. Check your state education department to see what has been done for students in your location.

Common Core: Mathematics

Common Core standards for mathematics are divided into two parts: "Standards for Mathematical Practice" and "Standards for Mathematical Content." The Mathematical Practice section focuses on developing depth of conceptual knowledge and the practical use of mathematical concepts. The Mathematical Content section focuses on procedural knowledge: determining the appropriate steps to follow in order to formulate answers. Together these two parts promote mathematical understanding so that concepts can be integrated into practice both in math activities and in interdisciplinary work.

The Common Core identifies eight core standards for mathematics. Three of the eight core standards are of particular interest to integrating STEAM activities in the curriculum, and in correlating these standards with other disciplines. The full list is here with full descriptions for the three most widely applicable standards for STEAM. The remainder are listed in italics for context.

1. Make sense of problems and persevere in solving them (p. 6).

    Proficient students can determine the parameters of the problem; find solutions, considering alternative approaches; interpret trends; and use and create tables, graphs, and diagrams in the process.

2. *Reason abstractly and quantitatively.*

3. Construct viable arguments and critique the reasoning of others.

    Proficient students use assumption and definitions to construct logical arguments. They analyze and draw conclusions based on data. They are able to justify and communicate their findings to others. They can identify both valid and flawed elements in reasoning. They can decide whether what is presented makes sense and can pose pertinent question to clarify their understanding.

4. Model with mathematics.

Proficient students can use their mathematical understanding to solve problems in day-to-day activities and in society at large. They can apply what they know to simplify a complicated situation and be ready for revision as needed.

5. *Use tools appropriately.*

6. *Attend to precision.*

7. *Look for and make use of structure.*

8. *Look for and express regularity in repeated reasoning.* (National Governors Association Center for Best Practices, 2010, 6–7; ©Copyright 2010 National Governors Association Center for Best Practices and Council of Chief State School Officers. All rights reserved.)

Statement one, in its use of process for determining solutions, can align well with a library research model. Identifying projects that will promote both math and library instructions can build on this statement. Statement three goes to the analysis and synthesis librarians work for in interpreting information to create a product or solve a problem. The library can become a center for students to research and develop real-world solutions using mathematical concepts (statement four). These common goals can foster collaborative experiences to benefit students.

Common Core: Scientific and Technology Literacy

The Common Core standards for English Language Arts include a section on scientific and technology literacy, connecting literacy as a key element of science and technology learning. In this section, objectives around reading and analyzing text are outlined and discussed. These standards align well with research instruction provided in library settings, and with information literacy research models. There are several standard statements at the middle and high school grade bands that also apply to the hands-on approach promoted in STEAM education, giving a wider scope for use in promoting STEAM activities. Those grade band statements are listed here, along with their standard notation code:

Grades 6–8: Compare and contrast the information gained from experiments, simulations, video, or multimedia sources with that gained from reading a text on the same topic. CCSS.ELA-LITERACY.RST. 6-8.9 http://www.corestandards.org/ELA-Literacy/RST/6-8/

Grades 9–10: Follow precisely a complex multistep procedure when carrying out experiments, taking measurements, or performing technical tasks, attending to special cases or exceptions defined in the text. CCSS.ELA-LITERACY.RST.9-10.3. http://www.corestandards.org/ELA-Literacy/RST/9-10/

Grades 11–12: Synthesize information from a range of sources (e.g., texts, experiments, simulations) into a coherent understanding of a process, phenomenon, or concept, resolving conflicting information when possible. CCSS.ELA-LITERACY.RST.11-12.9. http://www.corestandards.org/ELA-Literacy/RST/11-12/ (©Copyright 2010 National Governors Association Center for Best Practices and Council of Chief State School Officers. All rights reserved.)

In each of these statements, there is a connection between reading materials and the actual process of creating a product. Each can be used to make connections between library resources in digital or print form and the projects classroom teachers are crafting with students. This would build from conceptual knowledge to practical application as instruction is planned. As librarians collaborate with classroom teachers for instruction, use these statements as a way of connecting to classroom work.

For both Common Core areas, the American Association of School Librarians has produced crosswalk documents between these standards and those of the Association's *Standards for 21st-Century Learners*. That correlation can be impactful for working with classroom teachers and administrators. It can help show how libraries support curricula and build connections for activities and collaboration.

Next-Generation Science Standards

The current iteration of national science standards for K–12 education moves away from looking only at science concepts. They now take a more expansive view of science as a discipline that informs society at large, taking stock of issues that affect daily life. The standards developers made a conscious effort to see every child as a scientist who will continuously learn about science outside of school and in the workplace. Engineering concepts were embraced as an applied use of scientific principles for practical outcomes. Key to the new approach was to extend understanding of the sciences in three dimensions: scientific and engineering practices, crosscutting concepts that unify the study of science and engineering, and core ideas in natural sciences (Committee on a Conceptual Framework for New K–12 Science Education Standards, 2012, p. 10). Repetition of these concepts across grade levels and across disciplines within science helps students to better understand them. In this way, familiarity increases, and learning spirals learning as new and more sophisticated concepts are added, building and expanding student knowledge.

In working through the science standards, librarians have an added resource from the American Association of School Librarians. It is a standards correlation document, available online and listed at the end of this chapter. In this K–12 correlation document, middle school correlation begins on page 84 and high school at page 102. As an example of the content, on page 98 of this document, middle school science standards for earth and space sciences include this standard: "MS-ESS3-5. Ask questions to clarify evidence of the factors that have caused the rise in global temperatures over the past century." There are a number of AASL standard statements that would correlate with this content standard, including: "1.1.1 Follow an inquiry-based process in seeking knowledge in curricular subjects, and make the real-world connection for using this process in own life" and "1.2.1 Display initiative and engagement by posing questions and investigating the answers beyond the collection of superficial facts" (American Association of School Librarians, Correlations, 2015).

In the science standards, the engineering design section includes three stages: defining the problem, developing possible solutions, and improving designs. The standards address using these stages in all areas of science. The approach is to use the stages to address global issues. In defining the problem, students analyze an issue using both quantitative and qualitative information. In developing solutions, students would break the issue into more manageable pieces that can be addressed with engineering solutions, determining criteria, and prioritizing actions. In improving design, students use methods to model proposed solutions, taking into account a range of criteria and constraints and examining societal and environmental impacts as they test their products (NGSS Lead States, DCI Arrangement of the Next Generation Science Standards, 2013, 81). Given the role librarians play in research, this area of the standards is a natural connection to current efforts. Building more hands-on ways to demonstrate knowledge will help support the work of classroom teachers.

Standards for Technological Literacy

The International Technology and Engineering Educators Association (ITEEA) standards were developed by an association of teachers traditionally linked to the "industrial technology" field, the "shop teachers" of the world. As this group evolved, they redefined themselves as instructors of the designed world, introducing students to building, making, and shaping devices, structures and procedures to improve the material world. Among the educational standards for young people, ITEAA's document is the only one that explicitly builds standards around engineering concepts and processes. The ITEEA approach considers technology education the study of the human-made world. This is in contrast to science, which studies the natural world. ITEEA also contrasts their mission with that of educational technology, where tools are used to advance learning and teaching.

A core building block of this practical application of science and mathematical concepts is the engineering design process. As stated in the standards:

In order to comprehend the attributes of design, students in Grades 9–12 should learn that

> H. The design process includes defining a problem, brainstorming, researching and generating ideas, identifying criteria and specifying constraints, exploring possibilities, selecting an approach, developing a design proposal, making a model or prototype, testing and evaluating the design using specifications, refining the design, creating or making it, and communicating processes and results. The design process is a systematic, iterative approach to problem solving that promotes innovation and yield design solutions. To systematically seek an optimum design solution, engineers and other design professionals use experience, education, established design principles, creative intuition, imagination, and culturally specific requirements. (ITEEA, *Standards*, p. 97; International Technology and Engineering Education Association. Reprinted with permission. www.iteea.org)

The ITEEA standards align well with the research process, in that students research available information and principles to determine what works to solve a design problem, requiring the analysis and synthesis phases of the research cycle. Because this is a discipline that places a high premium on hands-on activities, library efforts may need to be redoubled to encourage collaboration that could take students out of the classroom setting. Teachers in this area would be very helpful in establishing makerspaces both as an extension of what happens in the technology classroom, and as a lure for students who may not have considered formal classes in this area.

ISTE Standards for Students

The International Society for Technology in Education (ISTE) standards are intended to be used to embed technology skills in all subject areas. A premium is placed on students taking ownership of their learning, which can play out nicely in library-related activities, as well as in the development of interest in science and technology career priorities for students. There are common themes within the standards that parallel many of the STEAM subject disciplines. ISTE identifies seven characteristics of learners and develops indicators for each. In this way an "innovative designer" (ISTE Standards for Students, 2016) can be matched with engineering-related standard statements in the *Next Generation Science Standards*. The Common Core standards focus on analysis and communication as well as content knowledge, so the ISTE standards can underlay work in the sciences and mathematics as well.

The ISTE standards include areas that dovetail well with AASL's *Standards for the 21st-Century Learners* so that both can support communication, collaboration, and problem-solving. AASL highlights the ethical use of resources, which mirrors the digital citizenship indicators in the ISTE Standards. These areas of common focus align so well that there is a natural coalition to be built between technology and library. Both the technologist and the librarian can take leadership roles in both instruction and collaboration with other educators.

National Core Arts Standards

Arts educators have come together to create a national set of standards for art education in five areas: dance, media, music, theater, and visual arts. The *National Core Art Standards* codify what represents success and achievement in the arts, focused on fundamental creative practices: imagination, investigation, construction, and reflection.

Within these areas, there are anchor standards grouped in four areas: creating, performing/presenting/producing, responding, and connecting. This standards list resembles the design process, as it fosters the steps of generating ideas, developing a project, and synthesizing the elements into a product that can be shared with others. In its conceptual framework there is specific mention that these creative practices have a bearing on science and mathematics (*National Core Art Standards*, 2014).

How will this play out in the school curriculum and in collaborative learning efforts? The arts can add creativity to projects as well as offering new avenues for communicating what has been learned. Scientific concepts can certainly be portrayed in art, dance, or music, and mathematical concepts inform art and design.

Library standards correlate well with the core ideas, which reflect the research model. In terms of library connections, we look to analyze and synthesize information. In the arts, there is also synthesis between the art form and the themes that are being explored. While libraries have traditionally looked at formal papers in research, consider the arts for projects that build on arts standards and on the kinesthetic and artistic intelligences of our young people. Because the library serves all curricular areas, the librarian can bring together seemingly disparate partners to develop interdisciplinary learning experiences for students.

Principles and Standards for School Mathematics

The National Council of Teachers of Mathematics (NCTM) standards document was published in 2000, a predecessor to the Common Core Mathematics document. It focuses on mathematical concepts and practices in the areas of number and operations, algebra, geometry, measurement, data analysis, and probability. It also addresses the processes of problem-solving, reasoning and proof, communication, connections, and representation (National Council of Teachers of Mathematics, "Principles, Standards and Expectations," 2000). Of these areas, the first three areas parallel the librarian's focus on problem-solving, critical thinking, and communication. While it may seem that math teachers are resistant to library research, there is room to collaborate on activities that can illuminate their classroom work. Start a discussion with a receptive teacher and learn what instructional issues are of most concern. It could start a good brainstorming session about math and the library.

USING STANDARDS IN WORKING WITH SUPERVISORS AND ADMINISTRATORS

All educators want to prove value to their learning community. This is particularly true for librarians, who work with all disciplines and educators in all fields. To prove value, these factors need consideration: following the generally accepted methods and procedures used by a professional community, knowing what to measure, and understanding the concepts and parameters of STEAM classes and programming. Professional practice is stronger when using existing standards, and when having a broad view of the intellectual and developmental needs of young people. Being conversant in the standards for all subject disciplines touches all areas of librarianship: working with students, selecting materials, designing instruction, building programs, and evaluating programming and instruction.

In practical terms, the library program has a stronger position when the librarian can show how programming and instruction connects to the standards that students are being assessed on in subject disciplines. In mathematics and science, this extends to the role that librarians have in fostering skills that may be part of larger state and national assessments. While it is difficult to make direct connections in this area, being able to provide correlations between library instruction and the skills being met in content areas will show the library's role in the larger school community. For supervisors and administrators who may have a more limited view of the librarian's role, this can be eye opening. This can be self-reported by library staff, but having that work validated by teachers in subject areas is also most helpful.

In reporting out about research projects and activities in the library, connect a standard, just as classroom teachers cite the standard they are teaching for each lesson. This builds credibility as an instructional leader in the school and district. These connections can also provide a rationale for funding to help students, whether in purchasing materials for hands-on activities, developing the collection, or continuing provision of electronic databases for a particular field.

Another area to consider as an instructional leader is professional development, discussed in greater depth in a separate chapter. Use your knowledge of a variety of standards to provide professional development on the cross-disciplinary approaches to research, design, and inquiry. Since STEM and STEAM initiatives are often written in to school goals or strategic plans, a correlation can be made between classroom and library instruction.

As an extension of the collection development function of the library, make materials available to administrators, colleagues, students, and families both in hand and online. Provide displays and handouts to share with students and families about the connections between different areas of study under the STEAM umbrella, with a focus on library activities. Set up activity centers or a makerspace for informal STEAM experiences. Unlike classroom teachers, librarians rarely have the opportunity to interact with families. Share information with classroom teachers that they can hand out to families at curriculum nights and conference times. In all cases, highlight library standards that are subject-agnostic and build connections between different disciplines. Show how you interact with all facets of the school or library program.

Standards provide information on what young people should know and be able to do. They can also be the stepping-stones to connecting with other educators to create learning experiences for students. As STEAM initiatives grow in a school and district, standards can guide the growth and the understanding of commonalities between subject disciplines. Librarians can build connections for the library program and for continuing connections with other educators that will benefit young people.

RESOURCES
Standards Documents

American Association of School Librarians. *Standards for the 21st-Century Learner.* Chicago: American Association for School Librarians, 2007.
Note: Currently under revision, with new standards to debut in November 2017 at the biennial conference. Update information can be found at: http://www.ala.org/aasl/standards#standards

International Society for Technology in Education. "ISTE Standards for Students." 2016. https://www.iste.org/standards/standards/for-students-2016

International Technology and Engineering Educators Association. *Standards for Technological Literacy: Content for the Study of Technology.* 3rd edition. 2007. https://www.iteea.org/File.aspx?id=67767&v=b26b7852

National Coalition for Arts Standards. *National Core Arts Standards; Dance, Media Arts, Music, Theater and Visual Arts.* 2014. http://www.nationalartsstandards.org

National Council of Teachers of Mathematics. *Principles and Standards for School Mathematics.* Reston, VA: National Council of Teachers of Mathematics, 2000.

National Council of Teachers of Mathematics. *Principles and Standards for School Mathematics, Grades 6–8 Edition.* Reston, VA: National Council of Teachers of Mathematics, 2000.

National Council of Teachers of Mathematics. *Principles and Standards for School Mathematics, Grades 9–12 Edition.* Reston, VA: National Council of Teachers of Mathematics, 2000.

National Governors Association Center for Best Practices, Council of Chief State School Officers. *Common Core State Standards.* Washington, DC: National Governors Association Center for Best Practices, Council of Chief State School Officers, 2010. http://www.corestandards.org

NGSS Lead States. *Next Generation Science Standards: For States, By States.* Washington, DC: The National Academies Press, 2013.

RELATED DOCUMENTS

American Association of School Librarians. *Learning Standards and Common Core State Standards Crosswalk.* 2011. http://www.ala.org/aasl/standards/crosswalk

American Association of School Librarians. Correlations between the AASL *Standards for the 21st-Century Learner* and the Next Generation Science Standards. 2015. http://www.ala.org/aasl/sites/ala.org.aasl/files/content/guidelinesandstandards/ngss/NextGen_AASL_by_Grade.pdf

American Association of School Librarians. *Empowering Learners; Guidelines for School Library Programs.* Chicago: American Library Association, 2009.

American Association of School Librarians. *Standards for the 21st-Century Learner in Action.* Chicago: American Library Association, 2009.

Committee on a Conceptual Framework for New K–12 Science Education Standards. National Research Council. *A Framework for K–12 Science Education: Practices, Crosscutting Concepts and Core Ideas.* Washington, DC: National Academies Press, 2012. http://www.nap.edu/13165 (available for free download as a PDF)

International Technology and Engineering Educators Association. *Advancing Excellence in Technological Literacy: Student Assessment, Professional Development and Program Standards.* Reston, VA: International Technology Education Association, 2007.

Massachusetts Department of Elementary and Secondary Education. "Science and Technology/Engineering Education for All Students: A Vision." 2016. http://www.doe.mass.edu/frameworks/scitech/2016-04/Vision.pdf

Mayes, Robert, and Thomas R. Koballa Jr. "Exploring the Science Framework." *Science Teacher* 79, no. 9 (2012): 27–34.

National Coalition for Arts Standards. *A Conceptual Framework for Arts Learning.* State Education Agency Directors of Arts Education (SEADAE), 2013. http://www.nationalartsstandards.org/sites/default/files/ConceptualFramework 07-21-16.pdf

National Research Council (NRC). *A Framework for K–12 Science Education: Practices, Crosscutting Concepts, and Core Ideas.* Washington, DC: National Academies Press, 2012.

National Science Teachers Association. "NGSS@NSTA STEM Starts Here." http://ngss.nsta.org

NGSS Lead States. *DCI Arrangement of the Next Generation Science Standards.* Achieve, Inc., 2013.

Professional Values Task Force. Young Adult Services Association. *Core Professional Values for the Teen Services Profession.* Chicago: American Library Association, 2015.

Search Institute. *40 Developmental Assets for Adolescents,* 2007. http://www.search-institute.org/content/40-developmental-assets-adolescents-ages-12-18

Arts-related documents: professional organizations that formed the National Coalition for Arts Standards also have sections on their websites regarding the standards, focusing on each specific arts area with information for members.

American Alliance for Theatre & Education (AATE). National Standards. http://www.aate.com/national-standards

Educational Theatre Association (EdTA). Standards Resources. https://www.schooltheatre.org/advocacy/standardsresources

National Art Education Association (NAEA). National Visual Arts Standards. https://www.arteducators.org/learn-tools/national-visual-arts-standards
National Association for Music Education (NAfME). Standards. http://www.nafme.org/my-classroom/standards/
National Dance Education Organization (NDEO). National Core Arts Standards (NCAS) in Dance. http://www.ndeo.org/content.aspx?page_id=22&club_id=893257&module_id=159040

The following two references are based on the National Core Art Standards, and are designed as a clickable database to individual standard statements by grade level. ArtsEdge also provides a searchable database of lesson plans that can be searched concurrently by a specific arts area and other STEAM disciplines. For instance, when searching media arts and science, one resource will be a lesson on the art of fireworks that aligns with the chemical and explosive properties of fireworks.

ArtsEdge. Standards for the Performing and Visual Arts for Grades 5–8. https://artsedge.kennedy-center.org/educators/standards/full-text/5-8-standards
ArtsEdge. Standards for the Performing and Visual Arts for Grades 9–12. https://artsedge.kennedy-center.org/educators/standards/full-text/9-12-standards

REFERENCES

American Association of School Librarians. Correlations between the AASL Standards for the 21st-Century Learner and the Next Generation Science Standards. 2015. http://www.ala.org/aasl/sites/ala.org.aasl/files/content/guidelinesandstandards/ngss/NextGen_AASL_by_Grade.pdf
American Association of School Librarians. *Learning Standards and Common Core State Standards Crosswalk.* 2011. http://www.ala.org/aasl/standards/crosswalk
Committee on a Conceptual Framework for New K–12 Science Education Standards. National Research Council. *A Framework for K–12 Science Education: Practices, Crosscutting Concepts and Core Ideas.* Washington, DC: National Academies Press, 2012.
Costa, Arthur. "Describing the Habits of Mind." In Arthur Costa and Bena Kalllick (eds.), *Learning and Leading with Habits of Mind.* Alexandria, VA: Association for Supervision and Curriculum Design, 2009. http://www.ascd.org/publications/books/108008/chapters/Describing-the-Habits-of-Mind.aspx
International Society for Technology in Education. "ISTE Standards for Students." 2016. https://www.iste.org/standards/standards/for-students-2016
International Technology and Engineering Educators Association. *Standards for Technological Literacy: Content for the Study of Technology.* 3rd ed. 2007. https://www.iteea.org/File.aspx?id=67767&v=b26b7852
National Coalition for Arts Standards. *National Core Arts Standards; Dance, Media Arts, Music, Theater and Visual Arts.* 2014. http://www.nationalartsstandards.org

National Council of Teachers of Mathematics. "Principles, Standards and Expectations." http://www.nctm.org/Standards-and-Positions/Principles-and-Standards/Principles,-Standards,-and-Expectations

National Governors Association Center for Best Practices, Council of Chief State School Officers. *Common Core State Standards*. Washington, DC: National Governors Association Center for Best Practices, Council of Chief State School Officers, 2010. http://www.corestandards.org

NGSS Lead States. *DCI Arrangement of the Next Generation Science Standards*. Achieve, Inc. 2013.

Professional Values Task Force. Young Adult Services Association. *Core Professional Values for the Teen Services Profession*. Chicago: American Library Association, 2015.

Search Institute. *40 Developmental Assets for Adolescents*, 2007. http://www.search-institute.org/content/40-developmental-assets-adolescents-ages-12-18

11

◇ ◇ ◇

EVALUATION AND MEASUREMENT

REPORTING YOUR SUCCESS

I loved the new Lego Blocks! The makerspace is amazing! The business database helped me finish my report, thanks! I loved the book you gave me; do you have any more like it?

Comments like this warm the hearts of librarians no matter what type of library they work in. To provide the best service for our patrons and our libraries we always strive to improve services and stay true to the mission and vision statements of our libraries. How do you do that? You set up an action plan to know what you want to accomplish. You follow your plan noting what worked and what did not, and you adjust as needed. You then reflect, analyze, and report on the successes and failures of your projects through evaluation and feedback. You determine if you've met the goals that you stressed in your action plan.

Action Plan

Gather the mission and vision statements of your library and the list of goals that you've set for your library. For example, if your goal is to plan a coding program what will be your objectives and how do you measure your success? Using the SMART goal acronym as a starting point helps to guide you through the process and to plan your evaluation tools. Smart goals put your needs into measurable and concrete terms; you'll find a thorough explanation by Duncan Haughey at https://www.projectsmart.co.uk/smart-goals.php.

S = Specific
M = Measurable
A = Attainable or Achievable
R = Realistic or Relevant
T = Timely or Time-based

Using the example of a collaborative coding program your objective might read as follows:

The Teen Librarian and Middle School Librarian will initiate two coding programs for seventh graders, using Hour of Code, during after-school sessions in October.

In this example you have **specified** those responsible for the program, as well as the program to be used. You have ensured that the program is **measurable** by detailing the number of sessions given and the audience. Your objective is **attainable** as it is open to only one grade level and deals with a finite number of programs at this stage. It is **relevant** as it allows young patrons to sample computer coding, a skill they may not receive during the school day. It is **time-based** as it will be accomplished in October. When your program is finished you can determine if you have achieved your goal—did you run two after-school sessions of Hour of Code in October for 7th graders? While this will not be the only measure of success, it is your starting point. Other measures of patron satisfaction including a survey after each session can be added. See another sample of SMART Goals in figure 1.1.

Add any information that you've gathered from a needs assessment, talk with staff and managers and as well as your Teen Advisory board. If you have no board of teens talk with your pre-teens and teens and discover their interests. Your regular visitors will likely have opinions on what needs to be added. Gather your research, from the experts you follow online via articles, presentations, TED Talks and podcasts, Twitter and Facebook, and other social media outlets.

Once you have some answers you can draft an Action Plan, the document that includes your goals, needs, budget, timeline, facility plan, potential experts on staff and in the community, and a plan for how you will staff the area and provide security. (See Figure 11.1 for a sample Action

Timeline	Year One	Year Two	Year Three
Goal	To create a makerspace in the library	To add engineering and robotics	To add video production and editing
Needs	Legos and crafts	Robotics	Green screen
Budget	Operating monies and donations	Grants + operating monies	Sustainable through increase in operating monies
Facility plan	Area	Enlarged area	Separate room
Partners—Experts	Parent volunteers Computer teacher	Engineering experts Math teachers	Local TV technicians
Partners—Financial	Educational foundation, Toy companies	STEM grants Engineering, computer firms Best Buy foundation grant	Local grants Innovation grants
Staff—Experts	IT department	Math teachers IT department	IT support
Staffing the area	Teen or school librarian	Teen or school librarian	Train an aide to supervise
Security for items	Storage bins, locked at night	Circulating only to classes or departments	Locked when not in use
Ages/limitations	Ages 12–15, plus Open House for parents and community	12–17, plus Open House for parents and community	12–17 plus Open House for parents and community

Figure 11.1: Action Plan

From *Full STEAM Ahead: Science, Technology, Engineering, Art, and Mathematics in Library Programs and Collections* by Cherie P. Pandora and Kathy Fredrick. Santa Barbara, CA: Libraries Unlimited. Copyright © 2017.

Plan.) Along with a budget, sustainability needs to be considered. How will you keep the space operational after the original funding is depleted? Will you receive a line item in your budget or need to seek grant monies? How do you provide a variety of activities to accommodate multiple intelligences and various ability levels? How do you ensure that we have activities that will appeal to both guys and gals?

Your goal statement must align with your institution's mission and vision statements. By aligning your projects with the adopted mission and goals of the school or public library you stand a better chance of securing buy-in from your management team. Most mission and vision statements are broadly written and that flexibility works to your advantage. Zero in on the goals for your department as well as the library's central goals. If one of the goals is to foster relations with the community you have a perfect reason to look for partners for any of your future projects.

Explain your needs in terms of your patrons. How will your proposal aid your patrons and/or the community? Will it change attitudes or behaviors? Will it provide a new skill set or open teens to the idea of STEAM careers? Will a makerspace allow pre-teens to solve real-world problems through hands-on activities?

Give thought to any facility needs in terms of space, maintenance costs, or staff coverage. Think also in terms of those outside your library who can advise you as you plan; list them as potential partners. Determine your needs in terms of materials. Do you have the materials you need on hand or will you need to build a budget? If operating monies will not stretch to cover the costs to whom will you appeal for funding? These factors all need to be in the budget portion of your action plan. Be sure that you have a rationale for the materials you need to buy and the budget that you propose. As you research items, collect bid sheets and save e-mails that pertain to cost factors. You may have to defend your intended purchase requests if you are asking your administrator for additional funds.

Evaluation

Just as you need an action plan for your programs you need an evaluation plan to measure effectiveness and decide about repeating or replicating it (Matthews, *Evaluation*, 2007, 8). School librarians and teachers are taught to create their tests before the lesson begins so that they meet standards and **know** which outcomes will be measured. Funding agencies want to know how you will measure the effectiveness of the program before you start the project. (See the Chapter 2 for further information on grants.) Determining what information you gather and analyze is a necessary first step to improving library services (Matthews, *Evaluation*, 2007, 8). Are you trying to encourage students to become more interested in STEAM subjects? Are you hoping to convince your manager to devote additional monies to your makerspace? Your goals will drive the evaluation tools that you use (Green, 2011, 89).

OUTCOMES-BASED EVALUATION

Outcomes-Based Evaluation or OBE is the term used for this process. OBE requires that you consider first what you want to achieve and then decide how to collect data that will measure your success. You are looking for ways to demonstrate how well you have met your goals. In turn this data helps you to show your library staff, your managers, and your supervisory boards how well you are serving your patrons and your community. The Institute of Museum and Library Services (IMLS) shares information in an article entitled "Outcome Based Evaluation Basics" (https://www.imls.gov/grants/outcome-based-evaluation/basics). Besides defining the term OBE, it provides examples under the headings of: program services, indicators, intended outcomes, and data sources. (These are located at the bottom of the page.) Two sets of examples are given, one for public libraries and one for museums; either set can be adapted to any library.

For examples of survey questions and focus group prompts turn to those provided by the California Library Association for its Summer Reading program. The page titled "Outcomes-based Summer Reading: Surveys and Focus Groups" located at http://www.cla-net.org/?page=84 provides samples of two surveys and two sets of prompts for focus groups. All are available in both English and Spanish. Another page entitled "Using Your Results" offers suggestions for reporting to your stakeholders and improving your own programs at http://www.cla-net.org/?page=343 (Janet Ingraham Dwyer, personal communication, March 23, 2017).

Outcomes-Based Planning

"I do think that librarians are well-advised to learn about outcome-based planning and evaluation, and to learn to evaluate programs in terms of outcomes—how the program fostered or inspired positive change in attendees' behavior/knowledge/etc."

Janet Ingraham Dwyer, Library Consultant, State Library of Ohio, Personal communication. March 23, 2017.

Wallace suggests that you first define your need and identify why you are undertaking the project (Wallace, 2001, 4–5). The Public Library Association's Project Outcomes Process includes steps to measure your success (or lack thereof), analyze your findings, and finally decide your next course of action (Davis, 2015, 35). Are you introducing students and staff to 3-D printing? Your own staff and your patrons need instruction in its use. Once they are familiar with the tools, you need to provide advanced training. Your programs will vary depending on the skill level and awareness of your teens and pre-teens. Identify the costs and materials and the amount of staff time needed (Wallace, 2001, 7).

There is a difference between evaluation and feedback. Evaluation is the process of reflecting on your project, measuring its impact, and determining which aspects should be repeated and which did not work as well as you hoped. Feedback is the information that you provide to your stakeholders after you've reflected on and evaluated the project. There are many ways to evaluate a program so don't feel that you have to use one set way of documenting your work. Keep in mind that there are many ways to evaluate your programs (Wallace, 2001, 7).

The process of evaluation requires two components: qualitative measures and quantitative measures. Qualitative measures are those that measure the human component and include observation and anecdotal records. Quantitative measures consist of counting and statistical measures: the number of lessons planned, the number of fantasy book purchased, the number of reader's advisory questions successfully answered are all quantitative outcomes (Everhart, 1998, 3). Librarians are always strapped for time so plan your evaluation tools to be as easy as possible so that they don't become an obstacle to the process (Matthews, *Evaluation*, 2007, 13). Your final decision is what you will do with the data collected (Wallace, 2001, 8).

Qualitative Measures

Qualitative measures allow you to collect stories of what worked, what didn't, and what excited your patrons (Markless, 2013, 153–54). Steve Jobs was a master storyteller in his corporate presentations. He believed you need to keep your presentations short and simple with no more than three main points as your goal (Gallo, Presentation Secrets, n.d.). (This short 11-slide presentation is no longer available online. Longer versions with similar titles are available by the same author.)

Storytelling allows you to seek opinions from your teen patrons, staff, other teachers, your managers, and any outside agencies that you work with as partners (Markless, 2013, 124). Did your patrons enjoy the program? Would they attend a similar program in the future? Did they tell their friends? Did usage increase? (Matthews, *Measuring*, 2004, 181) Were they happy with the new resources? Options for gathering qualitative measures include focus groups, interviews with participants, and simple observation by members of the library staff (Matthews, Joseph R., *Evaluation*, 2007, 47). One-minute reflection pieces are a powerful means of gathering the authentic voice of your teens (Pandora, 2013, 207). Ask an open-ended question and allow students to respond with words or drawings. You might ask, "After listening to the panel of female engineers what other speakers would you like to hear?" or "What one idea did you take away from their talk?" Feel free to add one or two multiple-choice questions as seen in the example below designed for middle school students.

Choose one answer for each question:

1. Was making the light saber card (http://raisinglittlesuperheroes.com/lightsaber-card/)

   too easy   easy   challenging   too difficult

2. After completing the green screen video which did you feel—a sense of

   accomplishment   or   frustration

   Help us to understand your answer by adding details here.

Customer Satisfaction Surveys can be designed to measure both qualitative and quantitative measures and can be done online via Survey monkey (www.surveymonkey.com), other free tools, or via a paper and pencil evaluation immediately following a workshop. The administrator in charge of school testing may have paid accounts to additional polling tools that you can use. Use a Likert scale to determine which programs to they'd like to see in the future (Matthews, Joseph R., *Evaluation*, 2007, 261). End the questionnaire by providing space for a short answer asking what they learned or what they want to learn next.

Your Likert scale can limit responses to three choices.

Do this again!
Never again (with an explanation of why not) and
Maybe but with these changes

Chances are when you started your project, program, or buying plan, you first analyzed the needs of your library. Your analysis of each program's success should refer back to the mission, and vision statements of both your department (e.g., YA department or school library) and of the public library or school district. Tweak the SWOT analysis tool (described in the Professional Development chapter) to gather comments from your pre-teen and teen patrons. They will likely be your most honest critics!

Have your students learned new skills? (Johnson, 2016, 58) Take pictures and videos to review and analyze later. (Some of these visuals will also be useful later for your reports.) Be aware that in many locations you cannot publish pictures of minors without parental permission (Markless, 2013, 129). One solution is to take the photos of student projects rather than of the students themselves. Another solution is to take pictures of the process by focusing on participants' hands without photographing faces. A third solution is to encourage teens to take pictures of themselves working; if they share these pictures with their friends you may find that you gather more participants. Reserve promoting the sharing of pictures on social media for your high school–age students. Even if participants are 18 or older, you need signed permission from them to publish a photo they appear in. If you use a photo taken by a student, you need the photographer's permission as well.

Quantitative Measures

Quantitative measures are those that involve numbers or relay numerical measures. One of the easiest quantitative measurements to gather is found in your budget—what specific items were ordered? (Pandora and Hayman, 2013, 62) How many items were purchased and what was the unit cost for each? What was the total cost of the project? You will need this information later for your managers but also for any agencies that funded your project.

Surveys are easy to analyze. Online surveys will tabulate your results for you (Matthews *Evaluation*, 2007, 13). You want patrons to be able to quickly answer your survey so keep it short and free from library terms not commonly known by patrons. With adults at professional development or technical training sessions it is easier to gain compliance due to credit incentives. The facilitator provides a website for feedback and the certificate detailing continuing education credit hours earned is e-mailed to them only *after* the survey is received and recorded.

Decide if you want your surveys to be anonymous. Anonymity provides for the most honest answers but hinders you from asking follow-up questions (Matthews, *Measuring*, 2004, 63). Survey options include checklists, short answers, or questions that require patrons to prioritize services (Mathews, *Measuring, 2004*, 64–66).

BADGING

Much of the work we do with young people is informal education. Assessment in these situations is often pointed toward the evaluation of the session rather than providing feedback to students about their achievement. Badging—a recognition of effort through standardized electronic badges—provides a way to show work in a given activity or workshop. As badges become recognized, they can signify accomplishment in a given area. Also known as micro-credentials, they can be used both for young people and for professional training for librarians and educators. This section is not intended to be a lengthy discussion of badging, but as an introduction of a tool that can be used in assessment and tracking of learning and participation.

How could a badging system be set up? Focus on competencies, using tools like the *40 Developmental Competencies for Adolescents* or the *Standards for the 21st-Century Learners*. Badges can be based on the skill areas listed in these documents. Another way to approach badging for STEAM activities, badges might be areas like

- Designer—use of design process in structuring a project
- Coder—exhibiting use of coding to create a project
- Builder—demonstrating construction of a product based on developing and following instructions

Badges as educational tools should be assigned consistently, and should represent actual effort rather than simple participation (Flickinger, 2016). They offer a way to recognize achievement in a visual format. There are open-source badging sites that allow users to design and display their own badges, where participants receive a code to redeem the badge online. If this is not an option for all young people, badges could be created as buttons, cards, and other formats. There are also badging sites that have pre-packaged badges, often involving a fee or subscription for their use.

Badging can also be used for professional development and become part of the librarian's portfolio. School librarians can also use them in the Master Teacher or licensure process. Think about badges in generic terms that they can be assigned for a variety of activities around the same competencies or concepts.

FEEDBACK

Feedback consists of communicating the results of your work to your stakeholders and doing so in a manner that meets the needs of your audience. You must decide what to do with the data collected (Wallace, 2001, 8). The most important idea to keep in mind is that you concentrate not on staff and materials but on your stakeholders; they are your bosses and can be the greatest cheerleaders that you have for your programs and institution (Gruenthal, 2012, 13). Analyze your results and describe how your activities have made a difference in the lives of your young adults (Davis, 2015, 33). Have your students learned skills in 3-D printing? In knitting? Have students produced artwork or shadowed an engineer? Are they enthusiastic about your offerings and asked for more? These are the types of metrics, defined as a standard of measurement, that you need to gather.

Teens should be considered stakeholders as well.

Discuss with your manager the best way to communicate with parents and community members. Does your library provide an online newsletter or is the library website the best means of communicating information? (Flowers, 1998, 206) Teens should be considered stakeholders as well and they might prefer an infographic or video on the website (Flowers, 1998, 207). Your communications serve three purposes—they provide information, position you as a leader, and allow you to drum up support so that you have advocates before the next levy or crisis comes (Gruenthal, 2012, 15). Note that not all information needs be shared with every group. There will be some information that should be kept in-house, that is, staff or personnel issues or difficulties uncovered during the process. This is an aspect to discuss with your managers. You don't want to over share problems with the community but want to try to solve them within your own organization.

Reports

There are numerous types of reports you can prepare. If preparing reports for grants you need both narrative and financial reports. To judge success you first want a user satisfaction type of report (Everhart, 1998, 6). Your board may want a formal annual report, if so, begin with an executive summary (Green, 2011, 90). In the narrative portion of your report enumerate your goals and objectives—the why and how of your project—explain the process and the types of statistics and anecdotal observations you've gathered. Your conclusion can detail any recommendations that you have for further programs, speakers, or activities as well as how you intend to fund the project in the future (Green, 2011, 90).

Narrative

Many funding agencies provide you with the required format for the final report so there is no guesswork involved. A narrative report is also required so stakeholders know your successes, difficulties, accomplishments, and how satisfied your users were with the project. If you already do an annual report in a written or visual format be sure to incorporate these items into your report. For an example of a visual annual report see Cougar Ridge Library, 2015–2016 by Julie Hembree (https://sway.com/yWLtk9YXpTTCyU9e).

Regardless of whether your report is for a grant agency or your board, be accurate. Report missteps and failures along with successes. Record your problems and suggest adjustments or solutions to improve your program. Format a simpler summary of your project and communicate it with your colleagues. They will appreciate learning from your trials and tribulations as well as from your successes. Include what you did not get to do as well as the changes you made during the duration of the project. Your managers will want to know if you believe the project proved to be valuable enough to warrant repeating it or if offering a more advanced version is your next step.

Financial Reports

Financial report requirements can be satisfied by a budget including bid sheets and an inventory of purchases reporting how the monies were expended. Include graphs designed from the statistics you create. A pie chart showing how money was spent will translate well into an annual report, social media, website, or presentations.

COMMUNICATION

Consider writing an article or making presentations to share your experiences. Other librarians benefit from knowing what worked, what didn't, and the pitfalls you encountered. Librarians believe that you shouldn't start from scratch if you can learn from other's experiences, so help them by sharing yours. Don't force others to "re-invent the wheel!" I was once told by a mentor that it was our "duty" to share with colleagues. So go forth and share!

RESOURCES—BADGING

Academy Badge Library. https://www.openbadgeacademy.com/mozillalearning/
 A variety of badges around information literacy include badges and activities for coding and design among a series focused on using information to create, change, and communicate on a website.
Badge Studio. http://www.badgestudio.com/badgeStudio/index.htm
 If budget permits, this vendor can produce badges to client instructions.
Credly. https://credly.com
 Both free and subscription pricing available to create badges
Open Badges. https://openbadges.org
 This open source site from Mozilla includes both prepackaged badges and a utility to create your own.

FURTHER READING—BADGING

Ferrari, Ahniwa. Badging the Library: Part 1. February 21, 2013. WebJunction. http://www.webjunction.org/news/webjunction/badging-the-library-part-01.html
Ferrari, Ahniwa. Badging the Library: Part 2. April 11, 2013. WebJunction. http://www.webjunction.org/news/webjunction/badging-the-library-part-02.html
Flickinger, Brad. *Reward Learning with Badges*. Arlington, VA: International Society for Technology in Education, 2016.
Fontichiaro, Kristin, and Angela Elkordy. "Chart Student Growth with Digital Badges." February 26, 2015. https://www.iste.org/explore/articleDetail?articleid=320
"What Is A Badge?" (video) 2:44. Mozilla Foundation. https://youtu.be/HgLLq7ybDtc

REFERENCES

Anthony, Carolyn A. "Project Outcomes—Looking Back, Looking Forward." Public Libraries Online. January 20, 2016. http://publiclibrariesonline.org/2016/01/project-outcome-looking-back-looking-forward/
California Library Association. "Outcomes-based Summer Reading: Surveys and Focus Groups." http://www.cla-net.org/?page=84
California Library Association. "Using Your Results." http://www.cla-net.org/?page=343
Davis, Denise, and Emily Plagman. "Project Outcome: Helping Libraries Capture Their Community Impact." *Public Libraries* 54, no. 4 (July/August 2015): 33–37.

Everhart, Nancy. *Evaluating the School Library Media Center: Analysis Techniques and Research Practices.* Santa Barbara, CA: Libraries Unlimited, 1998.

Flickinger, Brad. "5 Tips for Creating a Credible Badging Program." July 12, 2016. https://www.iste.org/explore/articleDetail?articleid=784&category=In-the-classroom&article=

Flowers, Helen F. *Public Relations for School Library Media Programs; 500 Ways to Influence People and Win Friends for Your School Library Media Center.* New York: ALA Neal-Schuman Publishers, 1998.

Gallo, Carmine. *The Presentation Secrets of Steve Jobs.* PowerPoint presentation, n.d.

Green, Ravonne A. *Case Study Research: A Program Evaluation Guide for Librarians.* Santa Barbara, CA: Libraries Unlimited, 2011.

Gruenthal, Heather. "A School Library Advocacy Alphabet." *Young Adult Library Services (YALSA)* 11, no. 1 (Fall 2012): 13–17.

Haughey, Duncan. "Smart Goals." ProjectSmart. 2017. https://www.projectsmart.co.uk/smart-goals.php

Hembree, Julie. "Cougar Ridge Library: A Year in Review 2015–2016." June 22, 2016. https://sway.com/yWLtk9YXpTTCyU9e (cited January 21, 2016)

Ingraham Dwyer, Janet. Personal communication. March 23, 2017.

Institute of Museum and Library Services (IMLS). "Outcome Based Evaluation Basics." https://www.imls.gov/grants/outcome-based-evaluation/basics.

Johnson, Abby. "Pulling the Plug." *American Libraries* 47, no. 5 (May 2016): 58.

Markless, Sharon, and David Streatfield. *Evaluating the Impact of Your Library.* 2nd ed. London: Facet Publishing, 2013.

Matthews, Joseph R. *The Evaluation and Measurement of Library Services.* Westport, CT: Libraries Unlimited, 2007.

Matthews, Joseph R. *Measuring for Results: The Dimensions of Public Library Effectiveness.* Westport, CT: Libraries Unlimited, 2004.

Pandora, Cherie P., and Stacey Hayman. *Better Serving Teens through School Library—Public Library Collaborations.* Santa Barbara, CA: Libraries Unlimited, 2013.

Raising Little SuperHeroes. "May the Fourth Be With You Light Saber Card." 2017. Blog. http://raisinglittlesuperheroes.com/lightsaber-card/

Wallace, Danny P. *Library Evaluation: A Casebook and Can-Do Guide.* Santa Barbara, CA: Libraries Unlimited, 2001.

FURTHER READING—FEEDBACK AND EVALUATION

ALA. 2014. Step 4—Evaluation and evidence. School library program health and wellness. http://www.ala.org/aasl/advocacy/tools/toolkits/health-wellness

Anthony, Carolyn A. "Moving Toward Outcomes." *Public Libraries* 53, no. 3 (May/June 2004): 5–7.

Farkas, Meredith. "Reuse, Recycle, Share." *American Libraries* 46, no. 1/2 (January/February 2015): 28.

Fontichiaro, Kristin, and Buffy Hamilton. "Undercurrents." *Knowledge Quest* 43, no. 1 (September/October 2014): 56–59.

Hernon, Peter, Robert E. Dugan, and Joseph R. Matthews. *Getting Started with Evaluation*. Chicago: ALA Neal-Schuman, 2014.

King, Kevin. "Creating Magical Moments." *Public Libraries* 54, no. 1 (January/February 2015): 32–33.

Snyder, Maureen M., and Janet Roche. "Road Map for Improvement Evaluating Library Media Programs." *Knowledge Quest* 37, no. 2 (November/December 2008): 22–27.

12

◇ ◇ ◇

PROFESSIONAL DEVELOPMENT

To offer the best STEAM experiences for young people—and to become the best in a chosen profession—requires lifelong learning. There are always new avenues to explore and new ways to expand a knowledge base, particularly in the STEAM arena. Whether it is a professional learning network (PLN) or formal face-to-face instruction, practice improves when individuals continually challenge themselves to grow. Professional growth helps adapt to new situations and incorporate new information. For the purposes of the librarian's role in STEAM efforts, professional development takes two directions:

1. Personal learning about STEAM and the programs, policies, and elements that can be part of a library program.
2. Work as an instructional leader, offering professional development in the workplace and professional activities.

This chapter provides information for both strands of professional development.

YOUR PROFESSIONAL LEARNING NETWORK

Lifelong learning is a fact of life for many Americans, as shown by a PEW Research Center survey on how people seek out learning activities in their personal lives and in their professional lives. In a 12-month period, people sought out resources like magazines, meetings, conventions, and online courses. For those in the workforce, these resources were used mainly for career-related reasons such as improving skills and meeting requirements for certification (Horrigan, 3). Whether it's in online or in face-to-face situations, people want to learn more about topics that are meaningful in their lives.

Professional development itself has become personal, as individuals can mine the wealth of online resources that are available to develop a professional learning network, or PLN for short. A PLN can be any combination of people who have common interests, members of a school department, grade-level educators, or a committee, or likeminded people online, or STEAM activities, and a local group that might include librarians, educators, and community members who have a common interest in promoting STEAM activities. There may already be PLNs in your school or library that might be receptive to learning more about STEAM. Successful teams often have some ground rules about meeting or discussion times, topics, and activities. You can choose all of the resources you need to keep you updated on trends and educate you on topics relevant to your needs. Learn about online resources posted on educational forums and social media, as well as interest groups that may make presentations at conferences.

Determine Your Needs: SWOT Analysis

Already have a list of topics for professional growth? If not, take time to consider what you need to improve practice. You can use SWOT analysis to clarify thinking. Often used as a part of strategic planning in a group setting, use it to crystalize thinking around any initiative, including a personal plan for professional growth. The "weakness" area may highlight areas to build into your plan. If you would like more information on SWOT analysis, go to http://inalj.com/?p=42148 for personal evaluation, and to http://guides.newman.baruch.cuny.edu/swot for an organizational overview.

Once the SWOT analysis is complete, use the responses to develop a plan. This can be a plan for you individually or for a group. If you have had a team looking at this, a session to review and plan would be in order. The sections on strengths and weaknesses can lead to a list of areas for future learning, as well as areas where expertise exists to offer training. The opportunities responses can be a springboard for an action plan to develop activities around a learning goal. The threats section can lead to conversations with those who can help remove obstacles, as well as examining ways to personally influence overcoming them.

On an individual level, having a PLN creates a self-awareness of lifelong learning. Making this PLN public will show your students or patrons a visible model of continuous learning and growth. Be proactive in the choices you make to learn more about STEAM topics that can enhance the library program, rather than being directed by a supervisor or administrator. Through PLN efforts, a network of contacts can be built to help you develop activities for students, patrons, and colleagues.

Developing a PLN (Professional Learning Network)

Professional Publications

Access an online database of journals through your library, or through a local public or academic library, to explore what these sources might be. For professional associations, there are often materials available only to members. This is a good time to "collaborate" with your colleagues, asking them to share their professional articles around STEAM. In addition to published journals, there are now numbers of digital journals for librarians that may have occasional articles on STEAM topics. Once you're registered with the journal, e-mail updates are usually provided. Check out titles like *Technology & Learning* and *District Administration*. When you find articles of interest, make them links on your website or library publications for others to use too.

PLN Possibilities: Organizing and Sharing

There are many avenues for expanding a PLN, limited only by the exploration you do and the time available to monitor and track information. Some are traditional sources, others involve exchange of ideas with others, both face-to-face and remotely. Here are some options.

Find *journal articles or books* on a STEAM topic and invite colleagues to form a discussion group to review and discuss reading. This can lead to more concrete efforts to build STEAM activities into classes and informal learning situations.

Peer teaching and sharing is a great way to learn from colleagues and form professional relationships that can set the stage for collaboration. Set up visits to each other's area (library, classroom, lab). Pair up with school and public librarians to observe activities. Follow up these observations with peer-to-peer discussions on the experience. The discussions should come after both colleagues have been observed, and could use the SWOT format as a structure. Get administrative buy-in for these activities, particularly if coverage is needed while observations are being done. Make sure to present this to superiors and provide a rationale on the benefits these experiences will bring to each participant and the institution at large, particularly benefits to the young people you serve.

Use *Twitter* to connect with others involved in STEM and STEAM, both individuals and organizations. The 140-character limit means it is easy to scan your feed and pick up on links that provide ideas and inspiration. Search terms like "STEM," "STEAM," the names of people who have written articles or blogs that you admire, associations and organizations with a STEAM focus, and topics like makerspaces, coding, robotics, etc. Click "follow" for those that seem interesting. Use the "like" function to collect links under your profile for easy retrieval.

Join *PLNs available online*. Many are free, and registration gives access to materials, discussions, and professional development opportunities. Discussion lists are usually a part of professional organizations. You can connect with others via e-mail, learning from their experiences and asking questions.

Follow *online blogs* that focus on STEM and STEAM. For example, Edutopia's and KQED's Mindshift both highlight these topics, in addition to other offerings on pedagogy and learning.

Conferences

One of the most important things to do to keep your skills sharp is to attend conferences and professional development activities either in person or via the web. Conferences are a highly motivating means of renewing your skill set related to libraries and to STEAM topics. Consider conferences at all levels—national, state, regional, and local—some of which will be kinder to the budget and perhaps more palatable to your institution. One incidental benefit of face-to-face conferences is meeting exhibitors and examining new resources.

Take advantage of virtual conferences, which can ameliorate the cost of travel and lodging. Some may be based on professional membership, which gives you access to session materials. Look for information on your state library organization's list serve and on social media. Librarians who have an ongoing relationship with science, computer, engineering, art, and math teachers can ask them to share information from their conferences that can be used in STEAM projects. Some of these national organizations include:

- American Library Association (ALA) and divisions focused on young people
  - American Association of School Librarians (AASL)
  - Association of Library Service to Children (ALSC)
  - Young Adult Library Services Division (YALSA)
- American Society for Engineering Education (ASEE, annual K–12 conference)
- Association for Supervision and Curriculum and Development (ASCD)
- Council of Chief State School Officers (CCSSO) under the auspices of the National Governors Association Center for Best Practices
- International Society for Technology in Education (ISTE)
- International Technology and Engineering Education Association (ITEEA)
- National Coalition for Arts Standards (NCAS)
- National Council of Teachers of Mathematics (NCTM)
- National Science Teachers Association (NSTA) has an annual conference and hosts an annual STEM Forum and Expo

The chart in Figure 12.1 identifies the STEAM focus you may find at conferences sponsored by these professional organizations.

Here is a sampling of other subject-specific conferences offered nationally and regionally that would include information on STEAM topics.

- Computers in Libraries. http://computersinlibraries.infotoday. com is sponsored by InfoToday, publisher of technology-related journals for librarians. The conference features trends in digital information and librarianship.
- Maker Ed Convening. http://makered.org/maker-educator-convening-2017 is held annually as a national conference for those interested in makerspaces.
- Remaking Education Days. http://remakelearningdays.org offers a series of visitation days to schools in southwest Pennsylvania and West Virginia.
- STEM Think Tank and Conference. http://stemefg.org/index .php/think-tank-and-conference is sponsored by the Center for STEM Education for Girls, with sessions in areas such as partnerships, STEAM, and entrepreneurship.

Conference Sources and STEAM Connections			
Discipline	**Source**	**Interests**	**Conferences & Workshops**
Arts	NCAS	Artistic expression	Artistic process using creativity for enhance creation
Common Core Mathematics	CCSSO	Problem solving, mathematical models	Usually part of math and science gatherings focused on process and active learning
Common Core Scientific & Technological Literacy	CCSSO	Design experimental process	Design models using scientific and technological concepts to create and communicate products
Education and technology	SXSWedu South by Southwest conference	Digital learning and technological developments	This annual conference features talks by innovators and experts sharing their experiences and concepts.
Libraries	American Association of School Librarians (AASL)	Teaching, research, databases, evaluating resources, digital citizenship	Biennial national conference; online workshops available on eCollab
Libraries	American Library Association (ALA)	Research and evaluation of resources	Annual conference (summer), online workshops available via ALA divisions
Libraries	Digital Public Library of America (DPLA)	Sponsors PLhackathons, art, culture, American heritage, data of science,	15 million items from over 2, 000 libraries, archives and museums 501c3 DPLA FEST held annually
Libraries	AASL	Information seeking	Research models using analysis and synthesis of information collected to create new meaning
Mathematics	NCTM	Problem solving	Mathematical models using analysis and mathematical concepts in solutions
Science	NSTA	Scientific Inquiry	Annual national conference and training online
Technology	FETC	Technology integration	Florida Education and Technology Conference
Technology	ISTE	Design Process, Technology integration	Annual national conference, online workshops (free and for fee)
Technology	ITEEA	Engineering design	Annual conference, and online workshops

Figure 12.1: Conference Sources and STEAM Connections

From *Full STEAM Ahead: Science, Technology, Engineering, Art, and Mathematics in Library Programs and Collections* by Cherie P. Pandora and Kathy Fredrick. Santa Barbara, CA: Libraries Unlimited. Copyright © 2017.

THE LIBRARIAN AS PROFESSIONAL DEVELOPER

The beauty of a PLN is that it can also be used to build a presence as a professional trainer or developer. Keep in mind that the librarian doesn't have to be the expert presenter. Begin as an informed learner.

Expert and Learner

Keep in mind that the librarian doesn't have to be the expert presenter. Begin as an informed learner.

Social media can help bridge the divide between personal learning and offering professional development for others. Capitalize on your learning journey. Finding good links about STEAM? Create a curation site that acts as your repository and guides others to good resources. Reading a lot about STEAM? Take time to reflect on a journal article, a blog page, or a learning management system, sharing it with colleagues, and opening it up for comments. In public libraries, an online presence through blogs and informational articles can serve the same purpose. Social media allows learning to be visible to the larger community, building an even wider PLN that can benefit a greater audience.

Determine the Needs of Your Staff

Begin by looking at institutional goals, strategic plans, and organizational initiatives. Many schools and libraries have already set goals around STEM and STEAM initiatives. Talk with those in your institution who handle professional development. Go in with a plan of what the library may be able to offer in terms of events.

In your particular situation, connect with colleagues and ask about their ideas for professional development. Look for ongoing efforts, and for people assigned to provide professional development opportunities in your institution. Offer your services and begin a dialogue about how STEAM professional development can fit into library, building, or district goals.

Creating PD Products

Demonstrate your learning by sharing your expertise. If this sounds daunting, think of the work you already do to promote the library program. What resources have you already promoted for your colleagues?

Think of it as just one step further to share your expertise in working with STEAM concepts. This can take many forms, including presentations and online resources.

The SWOT analysis you used for developing a PLN is a good tool for organizing experiences as well. A personal knowledge base can be strength if it is deep and nuanced. It can be a weakness if knowledge is at the beginner level. In that case, inviting local or regional experts can be a good start. The opportunity could be a professional development day, when administrators and supervisors are looking for people to offer sessions. When it comes to "threats," time is often the main barrier to offering face-to-face experiences.

Align your efforts with the goals of the district. You are not alone as a professional developer. People charged with overseeing professional development may have assistants who can help getting materials together and providing locations and refreshments. If they promote activities for the organization, your sessions can become a part of that process.

Capitalize on the knowledge of colleagues in STEAM areas. Here is where the connections you have made with educators in these fields will help. While they may be hesitant to present a full session, they could be part of a larger program to share expertise. Make it manageable for people who are already busy. This can also be the start of collaboration and interdisciplinary activities. In a public library setting, a starting point may be an overview of what STEAM is all about for those working with the public and with outreach.

Consider videotaping any PD event to post on the school, district, or public library website for future reference. Post any links or reference materials that were a part of the session. This will make learning available at the convenience of the learner.

Many schools use learning management systems. If this is available to you, set up a PD section on that platform where you can provide follow-up materials or the videos recorded. Use of this resource can be controlled by password access, where only those in the library, school, or district have access. Outside presenters may be more comfortable sharing materials and videos within this structure. Take a leading role in preserving and promoting this information as a part of collection development.

Since schools and libraries often have common goals, consider doing joint events where staff from both institutions can benefit. This presumes a working relationship is already in place. If not, make that a starting point to make connections between libraries within the school and within the community. If there is a community college or university in the area, extend the collaboration to include them as well. Professional development planning can foster collaboration between the institutions and promote the use of both by classroom educators.

Planning and Presenting

Think about the most successful training in which you were a participant. Keep in mind what worked best, and what you remember about the session, how content was prepared and presented, and the qualities of the presenter.

- Use active learning as much as possible. If it is a talk, find ways to intersperse opportunities for reflection or group discussion to break up the time.
- Training may be limited to before or after-school or workday times, which are not optimal for learning. Narrow the focus to make the session manageable and memorable. For online training, chunk the information into manageable segments with activities between sections.
- Space matters. Make sure the room being used is suitable for the activity and has space for breakout activities.
- If this is a session held before or after the workday, refreshments are always appreciated. They can set a positive note as the session begins.
- Offer proof of participation, via an established procedure in the school or library. Look at badging as another option. (More about this in Chapter 11.)

In planning the event, think strategically about what is to be accomplished. Opportunities to get time with colleagues are often limited, so the precious time you have needs to be rich in content that will be compelling. Consider these ideas:

1. Clearly state the objectives of the session. What should attendees expect to learn?
2. Do attendees need to do any preparation before attending the session? Make this clear.
3. Determine the flow of the session, and how to foster engagement.
4. Create a session evaluation. What criteria will be used? How will success be measured? How will results be used to inform future professional development sessions?

Document what the session will cover. Outline topics and the time allotted to cover them. Figure 12. 2 provides an example of a session presented

Coding Session			
Time	Activity	Presenter	Materials
2:00–2:10 *10 minutes*	Introduction: Goals & Outcomes	Librarian	Session handouts
2:10–2:40 *30 minutes*	Coding Overview	Technology Integrationist	Slide presentation
2:40–3:20 *40 minutes*	Lab activity: hands on coding exercise	Technology Integrationist & Librarian	Each attendee at workstation
3:30–3:50 *20 minutes*	Discussion of exercise & benefits for students—first reactions on post-it notes	Librarian leads (has someone to scribe discussion)	Post it notes and large paper to post notes
3:50–4:00 *10 minutes*	Conclusions & Evaluation	Librarian	Evaluation forms

Figure 12.2: PD Planning Form Sample

by the librarian and the building technology integrationist. It is a session introducing teachers to coding.

This can easily be adapted to provide an agenda for attendees. Adult learners want to know in advance what they are going to do and what will be expected of them. The times listed keep presenters on track, giving a sense of flow and purpose. Your colleagues' time is precious, so stick with the time allotted, starting on time. Note that while this example was for an educator or librarian session, it could easily have been for a parent or community group. Figure 12.3 is a blank version of the planning form for your use.

The most difficult point in planning a presentation may be choosing a topic for the session, among the many possibilities. If STEAM is already a part of your program, you may find a focus in new additions to the program. If your school or library is just beginning to adapt activities to a STEAM approach, more basic sessions may be in order. Here are some starter ideas for professional development sessions.

Make-and-take sessions, perfect for makerspaces and hands-on activities focusing on STEAM connections. These are good to build awareness, and learning about what students are doing in an informal setting. The library can be seen as leading the effort.

Learn together sessions, where a guest presenter (including students) walks the participants through the use of specific tools or applications. This approach could work for introductions to 3-D printing, coding,

Name of Session

Date & Time: Total Time:
Location:
Attendees:

	Time	Activity	Presenter	Materials
Introduction	Time span *# minutes*			
	Time span *# minutes*			
Body of Presentation	Time span *# minutes*			
	Time span *# minutes*			
	Time span *# minutes*			
Conclusion / Evaluation	Time span *# minutes*			
	Time span *# minutes*			

Figure 12.3: PD Planning Form

From *Full STEAM Ahead: Science, Technology, Engineering, Art, and Mathematics in Library Programs and Collections* by Cherie P. Pandora and Kathy Fredrick. Santa Barbara, CA: Libraries Unlimited. Copyright © 2017.

gaming, and robotics. This could be developed as a series, or singleton sessions as new tools are added to the program. Don't forget that students might fit the expert category. Topics might include career options, STEAM skills employers need, and new technologies, and applications.

Speakers can include vendors as new tools are added to the school or library. Many vendors offer professional development as a part of the services they offer, and are happy to tailor efforts to the time you have available.

With a focus on new tools like coding, gaming, and robotics, a focused group of educators likely to use the tools in STEAM activities would foster cooperation and interdisciplinary approaches. Involve science, mathematics, and arts in the activities.

Other Options to Share Professional Development

Presenting a session for fellow educators is just one possibility. Consider other ways to spread the word about STEAM through a variety of educational opportunities you develop. Make a presentation for trustees or school board members. Submit articles for publication in a newsletter, blog, or journal.

Making presentations at conferences can help you reflect on your practice: successes and failures, programming ideas, and interdisciplinary projects and activities. While it may not seem that your actions are new or unique, others will find your experience a great starting point for their work. You will also be able to network with others, building your professional learning network.

RESOURCES FOR PROFESSIONAL DEVELOPMENT

This resource list provides a beginning list of resources that are rich in information for your PLN and for planning professional development.

Professional Publications

Professional associations have print publications as well as online resources for members only. There may be some items that are free resources not tied to membership. Check their websites for more information.

Search online databases for relevant articles. Use database functions for saving articles, creating alerts, and other features to help automate getting new information. Google has some similar options. Listed below are several publications designed specifically for STEAM and others that follow trends in education, including STEAM.

Association for Supervision and Curriculum Development. http://www.ascd.org
Some articles from *Educational Leadership*, *Education Update* and blogs can be searched on this site, with books and professional development available for purchase.

District Administration. https://www.districtadministration.com
Sign up for a free subscription, with features on educational trends and new products in the education market.

Diversity in STEAM. http://diversityinsteam.com
This journal is available as a two-year (four issues) print subscription for $25.00, or digitally for $11.99. A sample issue can be viewed on the website.

STEAM Insight. http://kdrpr.com/steaminsight
Sign up for this monthly e-mail newsletter on the KDRPR (STEAM consultant) site. They provide links to STEAM PD events.

The Steam Journal. http://scholarship.claremont.edu/steam/
Claremont College hosts this academic, peer-reviewed journal around STEAM research, with free access to all articles.

STEAMEd Quarterly Digital Magazine. https://educationcloset.com/steamed-magazine/ Samples posted on the website, individual issues available at $2.99 an issue. A yearly subscription is available for $5.99.

Online Resources

Websites

Edutopia. https://www.edutopia.org
Edutopia focuses on hot topics in education including interdisciplinary programs and technology integration. The STEM to STEAM: Resources Toolkit can be found at https://www.edutopia.org/stem-to-steam-resources

Mindshift. https://ww2.kqed.org/mindshift
KQED follows trends in education, sharing educational research and practice via blogs, with resources for educators.

STEM to STEAM. http://stemtosteam.org
The Rhode Island School of Design (RISD) includes case studies about the connection between arts and traditional STEM disciplines.

Successful STEM Education. http://successfulstemeducation.org/resources
Find out about STEM programs with details about their operations.

Blogs and Podcasts

Blog posts can help you keep up with a specific topic by signing up for e-mail updates or RSS feeds. Here are several blogs that feature STEAM as a start to your collection.

Library Makers: *CraftLab*. (blog) http://librarymakers.blogspot.com/search/
    label/CraftLab
    The Madison (WI) Public Library features programs for teens from their
    CraftLab sessions.
The Show Me Librarian. (blog) http://showmelibrarian.blogspot.com
    Amy Koester is a public librarian in Illinois. Use the labels section to choose
    STEAM to see her postings about the projects she has designed.
STEM in Libraries. (blog) https://steminlibraries.com
STEM Learning Lab. (blog) http://stemlearninglab.com/blog/. Read about
    events and issues in STEM education from this Calgary lab.

Podcasts are serial programs online that can be accessed via apps like iTunes and Google Play. Search them by topic, and use a search engine to get "best podcast" information for any topic. For science-related podcasts, the Library of Congress has a resource called "Science Reference Guides; Podcasts Webcasts & Other Digital Media Media Files" (https://loc.gov/rr/scitech/SciRefGuides/podcasting.html).

Social Media

Resources like Facebook, Twitter, Pinterest, Instagram, and SnapChat can now serve as resources to learn and to share more about STEAM and STEM. Each tool can be searched by topic, so that connections can be made to whatever topics fit your professional needs. Make connections with experts in STEAM fields, and think about institutions like science museums and non-profit organizations that share information relevant to your needs. Here is a sampling of tools to use.

Facebook. https://www.facebook.com: In addition to personal pages, many individuals, organizations, and associations have pages to share information about their activities and related news. Use the search function to like and follow those who deal with STEAM topics. Try STEM Teaching Ideas and STEAM Education as a start. You can also find individual posts on STEM and STEAM topics in this search. Your school or library may already have a Facebook presence. If not, you may want to consider adding an organizational page where you can include other organizations and share the page with your colleagues.

Instagram. https://www.instagram.com: Use this photo-sharing app to get ideas and inspiration. As on other social media sites, a search on STEM will come back with results that may add to understanding and program development. The app can be downloaded to your phone from this site.

Pinterest. https://www.pinterest.com: Expecting to use Pinterest for hobbies, special events and recipes? There are great ideas for STEM and STEAM that can be pinned too. Free registration at the site gives access to create and comment on posts. Boards can be public or private, and you can also create a business account if you intend to use Pinterest with your community. One example: WeAreTeachers has an extensive list of pins about STEAM at https://www.pinterest.com/weareteachers/stemsteam-lessons-activities-and-ideas.

Use "The Teacher's Guide to Pinterest" at Edudemic (http://www.edudemic.com/guides/the-teachers-guide-to-pinterest/) to get instructions and tips for effective use of Pinterest as a professional tool.

Snapchat. https://www.snapchat.com: This relatively recent addition to social media used by organizations began as a way to post information for a brief period of time. It makes the tool more ephemeral, but it is a very popular app with young people. If you are not using Snapchat, it is worth considering if it would be a helpful resource. For more information, check these two resources:

1. Alfonzo, Paige. "Snapchat in the Library." *American Libraries*, November 1, 2016. https://americanlibrariesmagazine.org/2016/11/01/snapchat-in-the-library/
2. Moreau, Elise. "What Is Snapchat? An Intro to the Popular Ephemeral App." January 10, 2017. https://www.lifewire.com/what-is-snapchat-3485908

Twitter. https://twitter.com: Follow STEM/STEAM topics—no need to comment or post your own tweets unless or until you feel comfortable. Search for STEM education, STEAM education, or search names of people connected to STEAM efforts and follow their Twitter feeds. There are also some useful references to help in using Twitter:

1. O'Neal, Keith. Educators to Follow (on Twitter). https://www.linkedin.com/today/author/0_2x1yv0u0MJiOxC7EGNVekY?trk=prof-sm: O'Neal is an instructional designer at Lipscomb University (TN) who has created lists of educators to follow in each state, most of whom have an interest in educational technology and curriculum.
2. Schrock, Kathy. "Cure What Ails You: A Dose of Twitter for Every Day." *Kathy Schrock's Guide to Everything.* http://www.schrockguide.net/twitter-for-teachers.html: Find a wealth of resources here: explanations, videos, examples, and recommended reading for educators from a library and technology guru.
3. "Twitter Chats 101: A Step-by-Step Guide to Hosting or Joining a Twitter Chat." https://blog.bufferapp.com/twitter-chat-101: Ready to move on from mere following to interaction with online experts? This guide offers help and resources.

4. Whitby, Tom. "If Twitter Is Not PD, What Is It?" 2014. ASCD Edge. http://edge.ascd.org/blogpost/if-twitter-is-not-pd-what-is-it: A useful article if you need to justify the use of Twitter for professional development, and by extension, other social media tools.

YouTube. https://www.youtube.com: This video-sharing site can be used both to locate videos around STEAM topics for professional development and to upload local efforts. Use the "About" link to learn how to use YouTube features. Google provides a good help page at https://support.google.com/youtube/answer/2802327?hl=en.

Communities (Professional Learning Networks)

ASEE's (American Society for Engineering Education) eGFI (Engineering, Go for It!). http://teachers.egfi-k12.org: Resources for teachers, including sign-up for monthly newsletter (also one for students), workshops, lesson plans, and other materials.

ALA and its divisions offer members-only features like wikis, discussion lists, and online learning. AASL's online learning network is eCollab at http://www.ala.org/aasl/ecollab. YALSA has online learning opportunities at http://www.ala.org/yalsa/onlinelearning. ALSC also offers online learning at http://www.ala.org/alsc/edcareeers/profdevelopment/alscweb

edWeb.net. http://home.edweb.net: It is an online educator community where members can join a subject-specific community as well as attend free webinars on a variety of topics (with certificates for participation). Membership is free, and users can add a variety of groups to their profile, and a variety of community groups can be added to the member profile.

International Society for Technology in Education (ISTE). https://www.iste.org: This membership organization has for-fee professional development around technology integration, discussion groups, and an annual conference. Some resources are also available for non-members on the site.

Library 2.0. http://www.library20.com: With a free registration, users are able to join discussions and participate in online webinars and find resources to support work in libraries.

Teachers First. http://www.teachersfirst.com: Sponsored by The Source For Learning, an educational technology company, Teachers First is "for teachers, by teachers" with online forums, lessons, time saving tools, webinars, and an e-mail newsletter.

Conferences, Webinars, and Workshops

Resources in this section are offered at a national level. Remember to check your state's professional associations for their offerings, beginning with your library association. You may want to start a local group of interested educators to share their learning experiences as a part of a PLN.

AASL eCollab. http://www.ala.org/aasl/ecollab hosts webinars and collects professional materials from conferences for members to access. AASL's eAcademy hosts longer courses, typically four to six weeks that are fee-based.

ASEE PreK-12 Engineering Workshop http://teachers.egfi-k12.org/asee-prek-12-workshop is offered each summer, with practitioners discussing successful lessons and presenting models for learning in engineering and STEM areas.

Discovery Education STEM Resources. http://www.discoveryeducation.com/what-we-offer/stem/index.cfm. In addition to their library of streaming media, Discovery Education has for-fee resources for STEM professional development.

EdCamp. http://www.edcamp.org. Also labeled the "Un-conference," the EdCamp Foundation fosters participant-driven professional development days. Organizers promote the event, and then attendees suggest topics they want to cover or facilitate, with the most popular topics forming the sessions. Each participant is both learner and presenter in an informal discussion.

edWeb.net http://home.edweb.net. EdWeb.net offers a wide variety of webinars free of charge, presented by practitioners. The webinars are also archived and available to view.

ITEEA. https://www.iteea.org. The International Technology and Engineering Educators Association provides both online learning experiences and an annual national conference.

Library 2.0. http://www.library20.com. Online conferences, webinars, and learning groups are yours here, with all events are archived for later access.

Makered http://makered.org. The professional development activities are a part of the online resource library, with the intent of providing hands-on experiences for educators entering or continuing in STEM/STEAM education.

National Science Teachers Association STEM Forum and Expo. http://www.nsta.org/conferences/stem.aspx is held annually to highlight information from practitioners and vendor information about products to use in STEM/STEAM programs.

TeachThought. http://wegrowteachers.com. This group has PD for a fee, and they also have some free resources: a blog, a free sign-up e-mail newsletter, and the TeachThought podcast. There are some sessions about STEAM topics. It may be most valuable for discussions of effective professional development.

USA Science & Engineering Festival. http://www.usasciencefestival.org. Science Spark sponsors this national festival that brings students and educators together to share information about projects. Check out festival videos and the speaker clearinghouse of scientists.

PD Online: Resources to Incorporate Locally

Community for Advancing Discovery Research in Education. http://cadrek12.org. CADRE offers K–12 educators information about research, products, people, and projects helpful for STEM programming. Check the resource spotlight, "Online and Blended Professional Development," at http://cadrek12.org/online-blended-professional-development

Exploratorium (San Francisco) exploTV. Iron Science Teacher videos http://www.exploratorium.edu/tv/archive.php?cmd=keyword&keywordtext=iron%20science%20teacher

Johnson Education: Professional Development for Educators https://www.nasa.gov/offices/education/centers/johnson/educators/index.html NASA's Johnson Space Center Office of Education offers both online and face-to-face PD, with a database of lessons and videos to use in the classroom.

NEON (NASA Educators Online Network). http://neon.intronetworks.com/#. Free sign-up for an account, allows searching network to find collaborators for professional development, materials for class project or mentors.

NSF 2016 Video Showcase. http://stemforall2016.videohall.com. This site collects videos about best practices in STEM education. Use this database of videos about best practices in STEM education, searchable by grade, institution, state, and topic. Videos can be used for PD sessions and for collaborative work to develop STEM and STEAM lessons.

Planting the Seeds for a Diverse U.S. STEM Pipeline: A Compendium of Best Practice K-12 STEM Education Programs. 2010. http://eie.org/sites/default/files/bayer_compendium.pdf. Sponsored by the Bayer Corporation and housed on the Engineering is Elementary site (Boston Museum of Science), this document outlines high-quality practices for STEM education, with program descriptions and the criteria used to judge them.

School Library Journal. http://www.slj.com/. SLJ offers webinars periodically through the year, some free and some for a fee. Sign up on the website to be on a mailing list announcing upcoming events (and other e-mail newsletter options) at http://www.slj.com/forms/SLJsub.php

TED. "Talks by Brilliant Women in STEM." https://www.ted.com/playlists/253/11_ted_talks_by_brilliant_wome. This TED playlist brings together talks by women in STEM careers about aspects of their work and research. This searchable database of TED talks has other videos around STEAM topics as well.

REFERENCES

Horrigan, John. "Lifelong Learning and Technology." Pew Research Center. March, 2016. http://www.pewinternet.org/2016/03/22/lifelong-learning-and-technology

SWOT Analysis: Strengths, Weaknesses, Opportunities, and Threats. https://eclkc.ohs.acf.hhs.gov/hslc/tta-system/operations/mang-sys/planning/2SWOTAnalysisS.htm

FURTHER READING

Abilock, Debbie, Kristin Fontichiaro, and Violet H. Harada, editors. 2012. *Growing Schools; Librarians as Professional Developers.* Santa Barbara, CA: Libraries Unlimited.

Avery, Zanj K., and Edward M. Reeve. "Developing Effective STEM Professional Development Programs." *Journal Of Technology Education* 25, no. 1 (Fall 2013): 55–69.

Bowler, Leanne. "Creativity through 'Maker' Experiences and Design Thinking in the Education of Librarians." *Knowledge Quest* 42, no. 5 (May–June 2014): 58–61.

Dow, Mirah J. "Creating a STEM-literate Society." *Knowledge Quest* 42, no. 5 (May–June 2014): 14–18.

Graves, Colleen. "Crafting Professional Development for Maker Educators." Edutopia. September 26, 2016. https://www.edutopia.org/blog/crafting-professional-development-maker-educators-colleen-graves

Jolly, Ann. "STEM PD: Expert Advice." 02/03/2013. MiddleWeb (Blog) https://www.middleweb.com/5805/stem-pd-expert-advice/

Krug, Don, and Ashley Shaw. 2016. "Reconceptualizing ST®E(A)M(S) Education for Teacher Education." *Canadian Journal of Science, Mathematics & Technology Education* 16, no. 2 (2016): 183–200.

Schwartz, Katrina. Can Micro-credentials Create More Meaningful Professional Development for Teachers? *Mindshift.* KQED. February 15, 2017. https://ww2.kqed.org/mindshift/2017/02/15/can-micro-credentials-create-meaningful-professional-development-for-teachers/

Appendix A

ART MUSEUMS

United States and Canada	
African American Art Washington, D.C.	http://africa.si.edu Smithsonian
Art Gallery of Greater Victoria Victoria, British Columbia, Canada	http://aggv.ca
Art Gallery of Ontario Toronto, Ontario, Canada	http://www.ago.net
Art Institute of Chicago Chicago, IL	http://www.artic.edu/
Canada Centre for Architecture Ville-Marie, Quebec, Canada	http://cca.qc.ca/
Cleveland Museum of Art Cleveland, OH	http://www.clevelandart.org/
Cooper Hewitt Museum New York, NY	https://www.cooperhewitt.org
de Young Museum San Francisco, CA	https://deyoung.famsf.org

(*continued*)

United States and Canada	
Denver Art Museum Denver, CO	http://denverartmuseum.org
Design Exchange Toronto, Ontario, Canada	http://www.dx.org
Eric Carle Museum Amherst, MA	http://carlemuseum.org Picture book art of Eric Carle
Gardiner Museum of Ceramic Arts Toronto, Ontario, Canada	www.gardinermuseum.on.ca/
Harvard Art Museum Harvard University Cambridge, MA	www.harvardartmuseums.org
Hirshhorn Sculpture Garden Washington, D.C.	http://hirshhorn.si.edu/collection/ home/#collection=home
Isabella Stewart Gardner Museum Boston, MA	www.gardnermuseum.org
Mazza Museum University of Findlay Findlay, OH	http://www.mazzamuseum.org Picture Book Art
Metropolitan Museum of Art (MOMA) New York	http://www.metmuseum.org
Montreal Museum of Fine Arts Montreal, Quebec, Canada	http://www.mbam.qc.ca
Museum of Fine Arts Boston, MA	www.mfa.org
National Gallery of Art Washington, D.C.	http://www.nga.gov
National Museum of African American History and Culture Washington, D.C.	https://nmaahc.si.edu/ Smithsonian
National Museum of the American Indian Washington, D.C.	http://nmai.si.edu/ Smithsonian
National Portrait Gallery Washington, D.C.	http://www.npg.si.edu
Reynolds House Museum of American Art Winston-Salem, NC	http://reynoldahouse.org
Royal Ontario Museum (ROM) Toronto, Ontario, Canada	http://www.rom.on.ca/en/ collections-research
Smithsonian American Art Museum Washington, D.C.	http://americanart.si.edu/

Europe	
The Louvre Paris, France	http://www.louvre.fr/en/
Musee d'Orsay Paris, France	http://www.musee-orsay.fr/en/ collections/overview.html
Museo del Prado Madrid, Spain	https://www.museodelprado.es/ en/the-collection

Appendix B

SCIENCE AND TECHNOLOGY MUSEUMS AND PLANETARIUMS

Science and Mathematics	
British Columbia Aviation Museum, Sidney, British Columbia, Canada	https://www.bcam.net/
California Science Center Los Angeles, CA	https://californiasciencecenter.org/ Space Shuttle Endeavor
Canada Science & Technology Museum Ottawa, Ontario, Canada	http://cstmuseum.techno-science .ca/en/education/try-this-out-math-game-fun.php
Carnegie Science Center Pittsburgh, PA	http://www. carnegiesciencecenter .org
College Park Aviation Museum College Park, MD	http://www.collegeparkaviation museum.com/Home.htm
Exploratorium San Francisco, CA	https://www.exploratorium.edu
Montreal Science Center Montreal, Quebec, Canada	http://www.montrealsciencecentre .com/

(*continued*)

Science and Mathematics	
Museum of Science Boston, MA	https://www.mos.org Math Moves permanent exhibit https://www.mos.org/exhibits/ math-moves
Museum of Science and Industry Chicago, IL	http://www.msichicago.org/
National Museum of Mathematics New York, NY	http://momath.org/
Ontario Science Centre Toronto, Ontario, Canada	http://www.ontariosciencecentre.ca/
Oregon Museum of Science and Industry Portland, OR	http://www.omsi.edu/ 2 Rue de la Commune Ouest2
Air and Space	
Air Force Space and Missile Museum Cape Canaveral, FL	http://www.afspacemuseum.org/ displays/
Canadian Air and Space Museum (CASM) Ottawa, Ontario, Canada	http://www.casmuseum.techno- science.ca/en/index.php
Experimental Aircraft Association (EAA) Oshkosh, WI	https://www.eaa.org/eaa
International Women's Air & Space Museum Burke Lakefront Airport Cleveland, OH	http://iwasm.org/wp-blog/
Intrepid Sea, Air & Space Museum Complex Pier 86 New York, NY	www.intrepidmuseum.org/Scout Space Shuttle Enterprise
National Air & Space Museum Washington, DC	https://airandspace.si.edu/ Smithsonian
National Museum of the U.S. Air Force Wright-Patterson Air Force Base near Dayton, OH	www.nationalmuseum.af.mil/ Expansion/GrandOpening.aspx
Pacific Aviation Museum Pearl Harbor, Ford Island, Honolulu, HI	http://www.pacificaviationmuseum .org/
Steven F. Udvar-Hazy Center Air & Space Museum Parkway Chantilly, VA	www.airandspace.si.edu Shuttle Discovery
U.S. Space and Rocket Center Huntsville, AL	http://www.rocketcenter.com/

Natural History	
American Museum of Natural History NY, NY	http://www.amnh.org/ Hayden Planetarium
Field Museum of Natural History Chicago, IL	https://www.fieldmuseum.org/
National Museum of Natural History Washington, D.C.	https://www.si.edu/Museums/ natural-history-museum Smithsonian
Natural History Museum of Utah Salt Lake City, UT	https://nhmu.utah.edu/
Technology	
The Computer History Museum Mountain View, CA	www.computerhistory.org
National Museum of American History Washington, D.C.	https://www.si.edu/Exhibitions/ Places-of-Invention-4626 Places of Invention
The Tech Museum of Innovation San Jose, CA 95113	https://www.thetech.org/
Planetariums	
Adler Planetarium and Astronomy Museum	http://www.adlerplanetarium.org/ #6idx7fU2GUMFFEIp.97
Clark Planetarium Salt Lake City, UT	http://clarkplanetarium.org/
Halifax Planetarium Nova Scotia, Canada	http://astronomynovascotia.com/ index.php/planetarium

Appendix C

MAKERSPACE SUPPLY LIST

Item	Have	Need	Notes
Facility			
Movable tables			
Chairs			
Wall space			
Whiteboard			
Craft supplies			
Balsa wood			
Cardboard			
Chipboard			
Construction paper			
Crochet hooks			
Fabric			
Felt			
Game pieces			

(continued)

Item	Have	Need	Notes
Craft supplies			
Glitter			
Glue			
Glue gun			
Knitting needles			
Letters (stick on)			
Paper			
Paper cutter			
Pinking shears			
Ribbon			
Rubber cement			
Scissors			
Sewing machine			
Straws			
Thread			
Yarn			
Building			
Containers			
Cutting mats			
Duct tape			
Exacto knives			
Gears			
Hammer			
Kinex			
Magnets			
Nuts, bolts, screws			
Popsickle sticks			
PVC pipe			
Screwdriver			
Electronics			
3-D pen			
3-D printer			
3-D scanner			

Electronics			
Circuit board kits			
Computer			
Cubify	Esra Murray		
Design software	Esra Murray-Math		
Digital camera			
Green screen			
Microcontrollers			
Mobile devices			
Poster printer			
Printer			
Raspberry Pi	$35		
Robot kits			
Tinkercad	Esra Murray		
Name Brands			
Arduino	Circuitry microcontrollers		
Cricut machine	$250–300		
Little Bots			
Legos	Set projects		
Makey Makey	$50		
Silhouette machine	$250–300		
Snap Circuits	Basic $25		
Sphero			
Squishy Circuits	$25		
Stikbots			
Vex	Set projects		

Appendix D
VENDOR LIST

GENERAL

Ada Lovelace Day Education Pack. http://findingada.com/resources/resources-for-schools/. ADL provides materials including lesson plans, materials for women in STEM, posters, and coding information.

BrainPop. https://www.brainpop.com BrainPop includes some free resources, and the subscription product includes many learning games.

Maker Shed. https://www.makershed.com

SparkFun Electronics. https://www.sparkfun.com

STEMFinity. https://www.stemfinity.com
Choose a category and shop on this site. There is also a "free resource" link at the bottom of the homepage, with resources arranged by topic.

BOOKS

Cereri, Kathy. *Robotics: Discover the Science and Technology of the Future with 25 Projects* (Build It Yourself series). White River Junction, VT: Nomad Press, 2012.

Graves, Colleen and Aaron Graves. *The Big Book of Makerspace Projects: Inspiring Makers to Experiment, Create and Learn.* New York: McGraw-Hill Education, 2016.

Make magazine also publishes "Make: The Best of" books as a compendium of projects found in the magazine.

Martinez, Sylvia Libow, and Gary S. Stager. *Invent to Learn: Making, Tinkering and Engineering in the Classroom.* Torrance, CA: Constructing Modern Knowledge Press, 2013.

McLees, Trent, ed. *STEM and Making* (eBook only). Chicago: YALSA, 2017.

Mercer, Bobby. *The Robot Book: Build & Control 20 Electric Gizmos, Moving Machines, and Hacked Toys* (Science in Motion). Chicago: Chicago Review Press, 2014 (grades 4 and up).

3-D PRINTERS

Cube Pro. http://www.3dsystems.com/shop

Lulzbot Mini. https://www.lulzbot.com

MakerBot. https://www.makerbot.com

MakerGear. http://www.makergear.com

Polar Printer. http://www.polar3d.com

CODING / ELECTRONICS

Arduino (open-source electronics platform) https://www.arduino.cc
Arduino also offers a curriculum package recommended for interdisciplinary study: https://www.arduino.cc/en/Main/CTCprogram

Raspberry Pi. https://www.raspberrypi.org/
This foundation provides low-cost computers for learning. They develop free resources to learn more about computing and how to make things, along with training and free guides and videos. Check the "What is Raspberry Pi?" video at https://vimeo.com/raspberrypi

Scratch. https://scratch.mit.edu
Use this program for interactive stories, games, and animations. The "For Educators" page provides more tips and resources. Accounts are free, and teachers can also get an account that gives group controls.

TinkerCad. https://www.tinkercad.com
This online 3-D design and printing app is a free tool, with teacher resources.

See also the listing on the Computer Science Education Week section on third-party products: https://csedweek.org/educate/curriculum/3rd-party

ENGINEERING CURRICULA

CK-12. "Engineering: An Introduction for High School." http://www.ck12.org/book/Engineering%253A-An-Introduction-for-High-School/
This textbook is part of CK-12's FlexBook program, a collection of open-source texts that can be adapted by educators for use with students. Digital textbooks can be used by chapter or as a book.

ITEEA. Engineering by Design. https://www.iteea.org/STEMCenter/EbD.aspx
The International Technology and Engineering Educators Association has developed a standards-based model curriculum for grades K-12, providing courses of study for technological literacy. There are sample materials that can be accessed prior to purchase of course materials.

PLTW. https://www.pltw.org/faq
Project Lead the Way has developed curricula for elementary, middle, and high school students, presented online. There are fees for the programs and for participation, which includes access to a learning management system for the coursework and professional development for educators using the program. The FAQ page is a good starting point for an overview of PLTW.

GAMING

Breakout EDU. http://www.breakoutedu.com/. Provides kits for immersive game experiences, like escape rooms, complete with videos and setup information.

Kahoot! https://getkahoot.com. Create competitive games

Minecraft Education Edition. https://education.minecraft.net/

MAGAZINES/JOURNALS/NEWSLETTERS

eGIF News for Teachers (monthly e-mail subscription) http://teachers.egfi-k12.org
ASEE (American Society of Engineering Education) provides this newsletter as an e-mail format with a free signup on their teacher

resource page. The newsletter includes an activity, a feature on a person in a STEM related field, links to news and events.

Make Magazine

The STEAM Journal. http://scholarship.claremont.edu/steam/
This online publication from Claremont University features peer-reviewed research, project ideas, art samples, and reflections about the intersection of the arts and sciences.

MAKERS/MAKERSPACES

Kiwi Crate. http://www.kiwicrate.com
Packages an activity a month in a subscription-based product. Tinker Crate (science and engineering) and Doodle Crate (art and design) are options for 9–16-year-olds. The site has a page for schools regarding bulk pricing.

Lego Education Middle School. https://education.lego.com/en-us/middle-school/explore

Lego Mindstorms. https://www.lego.com/en-us/mindstorms

littleBits. http://littlebits.cc
Purchase electronics kits for use in makerspaces and classrooms. They provide educator content and an educator discount on purchases, which can be accessed with a free signup on the website. littleBits also has a Twitter account.

"Makerspace Resources: Articles, Links and More Resources to Help you on Your MakerEd Journey." Renovated Learning. http://renovatedlearning.com/makerspace-resources/

Makey-Makey. http://makeymakey.com.
Two collaborators at MIT developed invention kits to connect to computers for people to create projects. MM also has an educators section and an online community for users, where inventions can be posted, ideas shared, and questions asked, at Makey Makey Labz https://labz.makeymakey.com

ROBOTICS

Bee-Bot. https://www.bee-bot.us.
Designed for younger kids, but usable for middle and high schoolers too.

BirdBrain Technologies LLC. http://www.birdbraintechnologies.com
This company has robots used in computer learning experiences
around coding and robotics. Their robots include Finch and
Hummingbird, developed by researchers at Carnegie Mellon
University. The website also provides information about activities,
events, and curriculum.

Dash and Dot robot pack (with Tickle app). https://www.make
wonder.com

Ozobot. http://ozobot.com/stem-education

Sphero Education. http://www.sphero.com/education
Sphero has developed a number of robots that are popular in
education, including BB08, Ollie, and Sphero Sprk. The education
section of their website includes a blog, resources, manuals, and
funding ideas.

VIRTUAL REALITY

Google Cardboard. https://vr.google.com/cardboard

National Science Foundation. Resources for STEM Education. http://
www.nsfresources.org/home.html
Information on this site is from NSF-funded projects and includes
teacher development, instructional materials, assessments, and
research syntheses.

Oculus. https://www.oculus.com Creator of Rift and Touch

SketchUp. https://edshelf.com/tool/sketchup/
3-D modeling program available to K-12 institutions with state
educational grants. Check with state educational technology
agencies to determine whether licensing is available.

STEM By Design. https://www.stem-by-design.com
This is scientist and award-winning teacher Anne Jolly's website
with ideas for teachers for grades 4–8.

Successful STEM Education. http://successfulstemeducation.org
An initiative of the National Science Foundation, providing
information on quality practices and tools for STEM in K-12
education.

STATE INFORMATION

Colorado Dept. of Education Stem Resources page. https://www.cde
.state.co.us/stem/resources

National Governors Association (PDF) 2011. Building a Science, Technology, Engineering, and Math Education Agenda; An Update of
State Actions. http://www.nga.org/files/live/sites/NGA/files/
pdf/1112STEMGUIDE.PDF

New York State STEM Education Coalition. http://www.nysstemeduca
tion.org/

BIBLIOGRAPHY

Abilock, Debbie, Kristin Fontichiaro, and Violet H. Harada, eds. *Growing Schools; Librarians as Professional Developers*. Santa Barbara, CA: Libraries Unlimited, 2012.

Afterschool Alliance. "Computing and Engineering in Afterschool." *Afterschool Alert*. Issue Brief No. 62. 2013.

Alessio, Amy J., and Heather Booth. *Club Programs for Teens: 100 Activities for the Entire Year*. Chicago: American Library Association, 2015.

Alessio, Amy J, Katie LaMantia, and Emily Vinci. *A Year of Programs for Millennials and More*. Chicago: American Library Association, 2015.

American Association of School Librarians (AASL). *Correlations between the AASL Standards for the 21st-Century Learner and the Next Generation Science Standards*. 2015. http://www.ala.org/aasl/sites/ala.org.aasl/files/content/guidelinesandstandards/ngss/NextGen_AASL_by_Grade.pdf

American Association of School Librarians (AASL). *Empowering Learners; Guidelines for School Library Programs*. Chicago: American Library Association, 2009.

American Association of School Librarians (AASL). *Learning Standards and Common Core State Standards Crosswalk*. 2011. http://www.ala.org/aasl/standards/crosswalk

American Association of School Librarians (AASL). "Math Crosswalk." 2012. http://www.ala.org/aasl/sites/ala.org.aasl/files/content/guidelinesandstandards/commoncorecrosswalk/pdf/All_Math_Standards.pdf

American Association of School Librarians (AASL). "Reading Standards Literacy in Science/Technology." 2011. http://www.ala.org/aasl/sites/ala.org.aasl/files/content/guidelinesandstandards/commoncorecrosswalk/pdf/ReadingLitSciAllStandards.pdf

American Association of School Librarians (AASL). *Standards for the 21st-Century Learner*. 2007. http://www.ala.org/aasl/standards/learning

American Association of School Librarians (AASL). "Writing Standards for Literacy in History/Social Studies, Science, and Technical Subjects." http://www.ala.org/aasl/sites/ala.org.aasl/files/content/guidelinesandstandards/commoncorecrosswalk/pdf/WritingAllStandards.pdf

American Library Association (ALA). Office for Intellectual Freedom. "Workbook for Selection Policy Writing." 2017. http://www.ala.org/Template.cfm?Section=dealing&Template/ContenManagement/ContentDisplay.cfm&ContentID=11173

American Library Association (ALA). "Step 4—Evaluation and Evidence." School Library Program Health and Wellness. 2014. http://www.ala.org/aasl/advocacy/tools/toolkits/health-wellness

American Library Directory, 2013–2014. Medford, NJ: Information Today, 2013.

American Youth Policy Forum. "Understanding STEM Education: A Discussion of the Key Issues, Efforts and the Role of Federal Policy; A Special Briefing for Congressional Staff." January 27, 2012. http://www.aypf.org/wp-content/uploads/2012/05/STEM%20Facts%20and%20Resources%20Handout.pdf

Anthony, Carolyn A. "Moving Toward Outcomes." *Public Libraries* 53, no. 3 (May/June 2004): 5–7.

Anthony, Carolyn A. "Project Outcomes—Looking Back, Looking Forward." *Public Libraries* Online. January 20, 2016. http://publiclibrariesonline.org/2016/01/project-outcome-looking-back-looking-forward/

Ascione, Laura. "New Trump Laws Will Support Women in STEM Fields." *eSchoolnews TrumpEd*. March 27, 2017. http://www.eschoolnews.com/2017/03/27/new-trump-laws-will-support-women-stem-fields/

Avery, Zanj K., and Edward M. Reeve. "Developing Effective STEM Professional Development Programs." *Journal of Technology Education* 25, no. 1 (Fall 2013): 55–69.

Belser, Ann. "Exposing Students to STEM Fields Early Helps Girls, Minorities See Potential." *Pittsburgh Post-Gazette* (PA), November 17, 2015.

"Best Buy Grant Provides Mobile DigiLab at MCDL." *The Post*. December 5, 2015.

Boesdorfer, Sarah, and Scott Greenhalgh. "Make Room for Engineering; Strategies to Overcome Anxieties About Adding Engineering to Your Curriculum." *Science Teacher* 81, no. 9 (2014): 51–55.

Bomhold, Catharine, and Terri Elder. *Build It, Make It, Do It, Play It! Subject Access to the Best How-to-Guides for Children and Teens*. Santa Barbara, CA: Libraries Unlimited, 2014.

Booth, Heather, and Karen Jensen, eds. *The Whole Library Handbook: Teen Services*. Chicago: ALA, 2014.

Borman, Laurie D. "Makerspaces, Digital Literacy, Advocacy at AASL15." *American Libraries* 47, no. 1/2 (January/February 2016): 20.

Bowler, Leanne. "Creativity through 'Maker' Experiences and Design Thinking in the Education of Librarians." *Knowledge Quest* 42, no. 5 (May–June 2014): 58–61.

Braun, Linda W. "Everything Is Messy." *American Libraries* 46, no. 11/12 (November/ December 2015): 58.

Braun, Linda W., Maureen Hartman, Sandra Hughes-Hassell, and Kafi Kumasi. The *Future of Library Services for and with Teens: A Call to Action.* Chicago: American Library Association, YALSA, 2014.

Bruxvoort, Crystal, and James Jadrich. "Don't Short Circuit STEM Instruction; Exploring the Goals for Engineering and Science." *Science Teacher*, 83, no. 1 (2016): 23–28.

Bureau of Labor Statistics. John I. Jones. Beyond the Numbers. "An Overview of Employment and Wages in Science, Technology, Engineering, and Math (STEM) Groups." https://www.bls.gov/opub/btn/volume-3/an-overview-of-employment.htm

Bureau of Labor Statistics. "Most New Jobs," December 17, 2015, accessed January 21, 2016, http://www.bls.gov/ooh/most-new-jobs.htm.

Burke, John J. *Makerspaces: A Practical Guide for Librarians.* Lanham, MD: Rowman & Littlefield Publishers, 2014.

California Department of Education. News Release. "State Schools Chief Tom Torlakson Marks Women and Girls in Science, Technology, Engineering, and Mathematics (STEM) Week." April 6, 2015.

California Library Association. "Outcomes-based Summer Reading: Surveys and Focus Groups" http://www.cla-net.org/?page=84

Cano, Lesley M. *3D Printing: A Powerful new Curriculum Tool for Your School Library.* Santa Barbara, CA: Libraries Unlimited, 2015.

Cereri, Kathy. *Robotics: Discover the Science and Technology of the Future with 25 Projects* (Build It Yourself series). White River Junction, VT: Nomad Press, 2012.

Cho, Janet H. "Lego Thrives with Social Media Strategy." *The Plain Dealer.* (Cleveland, OH). September 9, 2016.

Citrin, James M. "What Parents Should Tell Their Kids About Finding a Career." *Harvard Business Review*, May 15, 2015, accessed January 21, 2016, https:// hbr.org/2015/05/what-parents-should-tell-their-kids-about-finding-a-career

Colby, Jennifer. "2,445 Hours of Code: What I Learned from Facilitating Hour of Code Events in High School Libraries." *Knowledge Quest* 43, no. 5 (2015): 12–17.

Coleman, Mary Catherine. "Design Thinking and the School Library." *Knowledge Quest* 44, no. 5 (May/June 2016): 62–68.

Collins, Cathy. "STEM and the School Library: A Marriage that Makes Sense." *Knowledge Quest.* Blog. April 29, 2016. http://knowledgequest.aasl.org/stem-school-library-marriage-makes-sense

Committee on a Conceptual Framework for New K-12 Science Education Standards. National Research Council. *A Framework for K-12 Science Education: Practices, Crosscutting Concepts and Core Ideas.* Washington, DC: National Academies Press, 2012. http://www.nap.edu/13165 (available for free download as a PDF)

Committee on STEM Education. "National Science and Technology Council." *Federal Science, Technology, Engineering and Mathematics (STEM) Education 5-Year Strategic Plan.* May 2013, accessed January 21, 2016, https://www.white house.gov/sites/default/files/microsites/ostp/stem_stratplan_2013.pdf

Cooksey, Ashley J. "Partnerships Beyond Four Walls." *American Libraries* 48, no. 1/2 (January/February 2017): 72.

Couri, Sarah. "Transforming Summers: Lessons from Public Libraries." *Knowledge Quest* 43, no. 5 (2015): 70–75.

Cunningham, Christine M. and Melissa Higgins. "Engineering FOR Everyone." 2014. *Educational Leadership* 72, no. 4 (2014): 42–47.

Cunningham, Christine. Museum of Science, Boston, personal communication, August 1, 2013.

Daniel, J., Abdun-Nabi, and Finkelstein, JJ. "Minority Students Are the Future of STEM." *Washington Post*, January 3, 2014.

Daugherty, Michael K. 2013. The Prospects of an "A" in STEM Education. *Journal of STEM Education*. 14, no. 2: 10–14.

Davis, Denise, and Emily Plagman. "Project Outcome: Helping Libraries Capture Their Community Impact." *Public Libraries* 54, no. 4 (2015): 33–37.

DeCoito, Isha. "STEM Education in Canada: A Knowledge Synthesis." *Canadian Journal of Science, Mathematics and Technology Education* 16, no 2 (2016): 114–128.

DeNisco, Alison. 2012. "Fab Labs: Using Technology to Make (Almost) Anything!" *District Administration*, 48, no. 11(2012): 34–37.

Doll, Carol A. *Collaboration and the School Library Media Specialist*. Lanham, MD: The Scarecrow Press, 2005.

Donald in Mathmagic Land. Walt Disney Productions, 1959. http://disney.wikia .com/wiki/Donald_in_Mathmagic_Land

Donovan, Lori. 2015. "Riding the STEAM Train: Planning from the District's Library Services Level." *Library Media Connection* 33, no. 4 (January/ February 2015): 6–8.

Dorrill, Shelley, and Jana Fine. "School and Public Library Cooperation: A Collaborative Conversation." *Knowledge Quest* 41, no. 4 (2013): 46–51.

Dougherty, Dale. "Make Education: Remembering Seymour Papert, Tool Guides for Kids, and More." http://makezine.com/2017/02/16/education/ February 16, 2017.

Dow, Mirah J. "Creating a STEM-literate Society" *Knowledge Quest* 42, no. 5 (May–June 2014): 14–18.

Editorial. "Effort to Spur Women, Minorities to Pursue STEM Careers Worthy." *Walla Walla Union-Bulletin (WA)*, March 23, 2016. *Newspaper Source*, EBSCOhost.

826 National, and Traig, Jennifer, ed. *STEM to Story: Enthralling and Effective Lesson Plans for Grades 5–8*. Hoboken, NJ: John Wiley and Sons, 2015.

Elkins, Aaron. "Let's Play!: Why School Librarians Should Embrace Gaming in the Library." *Knowledge Quest* 43, no. 5 (2015): 58–63.

"Engineering Design Process." (chart) Teach Engineering. https://www.teach engineering.org/k12engineering/designprocess

Engineering: Emphasizing the "E" in STEM Education (STEM Smart Brief). 2013. *CADRE*. http://cadrek12.org/sites/default/files/STEM%20Smart%20Engi neering%20Brief%20final.pdf

Evans, Julie. "Speak Up 2015 National Results: From Print to Pixel." http://www .tomorrow.org/speakup/from-print-to-pixel-may-2016.html

Everhart, Nancy. *Evaluating the School Library Media Center: Analysis Techniques and Research Practices*. Santa Barbara, CA: Libraries Unlimited, 1998.

Exner, Rich. "Minorities Now 38.4% of Nation." *The Plain Dealer*, June 24, 2016.

Farber, Matthew. "Students as Designers: Game Jams!" July 3, 2015. https://www
.edutopia.org/blgo/students-as-designers-game-jams-matthew-farber

Farkas, Meredith. "Making for STEM Success: Creating a Community of Tinkerers." *American Libraries* 46, no. 5 (2015): 27.

Farkas, Meredith. "Reuse, Recycle, Share." *American Libraries* 46, no. 1/2 (January/February 2015): 28.

Ferrari, Ahniwa. Badging the Library: Part 1. February 21, 2013. WebJunction. http://www.webjunction.org/news/webjunction/badging-the-library-part-01.html

Ferrari, Ahniwa. Badging the Library: Part 2. April 11, 2013. WebJunction. http://www.webjunction.org/news/webjunction/badging-the-library-part-02.html

Fiels, Keith Michael. "Libraries Transforming Communities." *American Libraries* 45, no. 5 (May 2014): 6.

Fink, Jennifer. "Crowdfunding the Classroom." *District Administration Online* (September 2016).

Fink, Megan P. *Teen Services 101: A Practical Guide for Busy Library Staff.* Chicago: YALSA, 2015.

Fleming, Laura. *Worlds of Making: Best Practices for Establishing a Makerspace for Your School.* Thousand Oaks, CA: Corwin, 2015.

Flickinger, Brad. "5 Tips for Creating a Credible Badging Program." July 12, 2016. https://www.iste.org/explore/articleDetail?articleid=784&category=In-the-classroom&article=

Flickinger, Brad. *Reward Learning with Badges.* Arlington, VA: International Society for Technology in Education, 2016.

Flowers, Helen F. *Public Relations for School Library Media Programs; 500 Ways to Influence People and Win Friends for Your School Library Media Center.* New York: Neal-Schuman Publishers, 1998.

Fontichiaro, Kristin, and Angela Elkordy. "Chart Student Growth with Digital Badges." February 26, 2015. https://www.iste.org/explore/articleDetail?articleid=320

Fontichiaro, Kristin, and Buffy Hamilton. "Undercurrents." *Knowledge Quest* 43, no. 1 (September/October 2014): 56–59.

Fostering Success of Ethnic and Racial Minorities in STEM the Role of Minority Serving Institutions. New York: Routledge, 2013.

Freiberger, Marianne, and Rachel Thomas. *Maths Squared: 100 Concepts You Should Know.* London: Apple Press, 2016.

Friday Institute for Educational Innovation. Middle School STEM Implementation Rubric. Raleigh, NC: Author, 2013. https://www.ncsmt.org/wp-content/uploads/2013/09/STEMAttributesRubric_MIDDLE_v4_Aug2013_v2.pd

Frye, Julie Marie, and Vaughn W. Nuest. "The Mountains are Calling and You Must Go: Spending Part of Your Summer at a National Park Service Site." *Knowledge Quest* 43, no. 5 (2015): 56.

Gallo, Carmine. *The Presentation Secrets of Steve Jobs.* PowerPoint presentation, n.d.

Gardner, Howard. *Multiple Intelligences; New Horizons.* New York: Basic Books, 2006.

The GCAA Makerspace. "Makerspace FAQ." January 13, 2017. https://gcaamaker space.wordpress.com/

Gerding, S., & MacKellar, P. "Grants for Libraries: A How-to-Do-It Manual and CD-ROM." New York: ALA, Neal-Schuman, 2006.

Gerlach, Jonathan. "STEM: Defying a Simple Definition." *NSTA Web Digest, NSTA Reports.* April 11, 2012, accessed January 21, 2016, http://www.nsta.org/publications/news/story.aspx?id=59305.

Gilbert, Amy, and Katherine Wade. "An Engineer Does What Now?" *Science Teacher* 81, no. 9 (December 1, 2014): 37–42

Gilliss, Apryl Flynn. "A Novel Idea." *American Libraries. American Libraries* 45, no. 5 (May 2014): 45–49.

Goerner, Phil. "Creating a School Library Makerspace: The Beginning of a Journey." *School Library Journal Online.* January 19, 2015. http://www.slj.com/2015/01/technology/creating-a-school-library-maker-space-the-beginning-of-a-journey-tech-tidbits/

Gonzalez, Heather B. and Jeffrey J. Kuenzi. "Science, Technology, Engineering, and Mathematics (STEM) Education: A Primer." Congressional Research Service, August 1, 2012. https://fas.org/sgp/crs/misc/R42642.pdf

Gorman, Christine. "How President-Elect Trump Views Science." *Scientific American.* November 9, 2016. https://www.scientificamerican.com/article/how-president-elect-trump-views-science/

Graves, Colleen, and Graves, Aaron. *The Big Book of Makerspace Projects.* New York: McGraw-Hill, 2016.

Graves, Colleen. "Crafting Professional Development for Maker Educators." Edutopia. September 26, 2016. https://www.edutopia.org/blog/crafting-professional-development-maker-educators-colleen-graves

Green, Ravonne A. *Case Study Research: A Program Evaluation Guide for Librarians.* Santa Barbara, CA: ABC-CLIO, Libraries Unlimited, 2011.

Gruenthal, Heather. "A School Library Advocacy Alphabet." *Young Adult Library Services (YALSA)* 11, no. 1 (Fall 2012): 13–17.

Gubnitskaia, Vera, and Carol Smallwood, eds. *How to STEM: Science, Technology, Engineering, and Math Education in Libraries.* Santa Barbara, CA: Libraries Unlimited, 2013.

Gura, Mark. *Getting Started with LEGO Robotics; A Guide for K-12 Educators.* Arlington, VA: International Society for Technology in Education, 2011.

Harrington, Eileen G. *Exploring Environmental Science with Children and Teens.* Chicago: American Library Association, 2014.

Hatch, Mark. *The Maker Movement Manifesto: Rules for Innovation in the New World of Crafters, Hackers, and Tinkerers.* Columbus, OH: McGraw-Hill Education, 2013.

Haughey, Duncan. "Smart Goals." ProjectSmart. 2017. https://www.projectsmart.co.uk/smart-goals.php

Hembree, Julie. "Cougar Ridge Library: A Year in Review 2015–2016." June 22, 2016. https://sway.com/yWLtk9YXpTTCyU9e (cited January 21, 2016)

Hennig, Nicole. *Apps for Librarians: Using the Best Mobile Technology to Educate, Create, and Engage.* Santa Barbara, CA: Libraries Unlimited, 2014.

Hernon, Peter, Robert E. Dugan, and Joseph R. Matthews. *Getting Started with Evaluation.* Chicago: American Library Association/Neal-Schuman, 2014.

Hinton, Marva. "Study Links After-School Programs to Improved STEM Knowledge." Blog. *Education Week.* March 1, 2017. http://blogs.edweek.org/edweek/time_and_learning/2017/03/new_study_examines_link_between_after-school_programs_stem_knowledge.html

Honey, Margaret, Greg Pearson, and Heidi Schweingruber, Editors. Committee on Integrated STEM Education; National Academy of Engineering; National Research Council. *STEM Integration in K-12; Status, Prospects and an Agenda for Research*. National Academy Press. 2014. http://www.nap.edu/read/18612/chapter/1

Hopwood, Jennifer L. *Best STEM Resources for NextGen Scientists: The Essential Selection and User's Guide*. Santa Barbara, CA: Libraries Unlimited, 2015.

Horrigan, John. "Lifelong Learning and Technology." Pew Research Center. March, 2016. http://www.pewinternet.org/2016/03/22/lifelong-learning-and-technology

Hough, Lauren. "Makerspaces in the Library." *Public Libraries*. 56, no. 2 (March/April 2014). http://www.search-institute.org/content/40-developmental-assets-adolescents-ages-12–18

Humes, Linda, and Lauren Newman. "Investigating the Crime of the 20th Century Using 21st Century Tools." *Knowledge Quest* 41, no. 4 (2013): 70–72.

Hynes, Morgan, Portsmore, Merredith, Dare, Emily, et al. "Infusing Engineering Design into High School STEM Courses." National Center for Engineering and Technology Education. 2011. http://ncete.org/flash/pdfs/Infusing_Engineering_Hynes.pdf

Hynes, Morgan, Portsmore, Merredith, Dare, Emily, et al. "Using Your Results" http://www.cla-net.org/?page=343.

Ingraham Dwyer, Janet. Personal communication. March 23, 2017.

Institute of Museum and Library Services (IMLS). "Outcome Based Evaluation Basics." https://www.imls.gov/grants/outcome-based-evaluation/basics

International Society for Technology in Education (ISTE) Connects. "What Should Be in Your Makerspace Toolbox?" March 8, 2016. https://www.iste.org/

International Society for Technology in Education (ISTE). ISTE Standards for Students, International Society for Technology in Education, 2016. https://www.iste.org/

International Society for Technology in Education (ISTE). "Redefining Learning in a Technology-Driven World." June 2016. http://www.iste.org/docs/Standards-Resources/iste-standards_students-2016_research-validity-report_final.pdf?sfvrsn=0.0680021527232122

International Technology and Engineering Educators Association (ITEEA). *Advancing Excellence in Technological Literacy: Student Assessment, Professional Development and Program Standards*. Reston, VA: International Technology Education Association. 2007.

International Technology and Engineering Educators Association (ITEEA). *Standards for Technological Literacy: Content for the Study of Technology*. 3rd ed., 2007. https://www.iteea.org/File.aspx?id=67767&v=b26b7852

Jacobson, Linda. "Coding's Finest Hour." *School Library Journal* 62, no.1 (January 2016): 11.

Jakes, David. "5 Considerations for Designing Makerspaces." Blog. September 16, 2016.

Jakubowicz, Collette. "Makerspaces Without a Space: Circulating Maker Kits for the School Library." Posted in *How to Be Brave, Makerspace?* December 21, 2014.

Jarrett, Kevin. "Makerspace Middle School Journey: Shop Class Rebooted . . . Digitally." *Edutopia*. Blog. August 5, 2015. https://www.edutopia.org/blog/making-makerspace-shop-class-rebooted-kevin-jarrett

Jensen, Karen. "Small Tech, Big Impact: Designing My Maker Space." February 1, 2016.

Jensen, Karen. "What's in Your Teen MakerSpace Manual? Forms Edition." SLJ: Teen Librarian Toolbox Blog, January 4, 2017. (MakerSpace Activity Planning Checklist and Program Planning Worksheet)

Jolly, Anne. "STEM PD: Expert Advice." February 3, 2013. MiddleWeb. Blog https://www.middleweb.com/5805/stem-pd-expert-advice/

Jolly, Anne. "STEM vs. STEAM: Do the Arts Belong?" *Education Week*, November 18, 2014, accessed January 21, 2016, http://www.edweek.org/tm/articles/2014/11/18/ctq-jolly-stem-vs-steam.html?tkn=YWND/0gO/NlW 6CUW48kknDeHqoF0t4aRuI5h

Johnson, Abby. "Pulling the Plug." *American Libraries* 47, no. 5 (May 2016): 58.

Johnson, Abby. "Reach Out through Outreach." *American Libraries* 45, no. 11/12 (November, December 2011):48.

Johnson, Larry, Samantha Adams Becker, V. Estrada, and A. Freeman. *NCM Horizon Report: 2014 K-12 Edition.* Austin, TX: The New Media Consortium, 2014. https://www.nmc.org/publication/nmc-horizon-report-2014-k-12-edition

Kent, Lauren. "5 Facts about American Students." *Pew Research Center Fact Tank. Pew Research Center*, August 10, 2015, accessed January 21, 2016, http://www.pewresearch.org/fact-tank/2015/08/10/5-facts-about-americas-students/.

Kepple, Sarah. *Library Robotics: Technology and English Language Arts Activities for ages 8–24.* Santa Barbara, CA: Libraries Unlimited, 2015.

Killeen, Erlene Bishop. "Supporting STEM to Remain Relevant." *Teacher Librarian* 43, no. 2 (December 2015): 52.

King, Kevin. "Creating Magical Moments." *Public Libraries* 54, no. 1 (January/February 2015): 32–33.

Knowledge Quest. STEM/STEAM Blog Archive. http://knowledgequest.aasl.org/category/blogs/stem/

Krug, Don, and Ashley Shaw. 2016. "Reconceptualizing ST®E(A)M(S) Education for Teacher Education." *Canadian Journal of Science, Mathematics & Technology Education* 16, no. 2 (2016): 183–200.

Kuhlmann, L. Meghann, Denise Agosto, Jonathan Pacheco Bell, and Anthony Bernier. "Learning from Librarians and Teens About YA Library Spaces." *Public Libraries* 53, no. 3(May/June 2014): 24–28. http://publiclibrariesonline.org/2014/07/learning-from-librarians-and-teens-about-ya-library-spaces

Kuzo, Joseph. "School Librarians: Key to Technology Integration." *Knowledge Quest* 44, no. 1 (2015): 74–76.

Lacey, Gary. "Get Students Excited—3D Printing Brings Designs to Life." *Tech Directions* 70, no. 22 (September 2010): 17–19.

Landgraf, Greg. "Making Room for Informal Learning." *American Libraries* 46, no. 3–4 (2014): 32–34.

Larson, Jeannette. CREW: A Weeding Manual for Modern Libraries: Texas State Library and Archives Commission, Austin, Texas, 2012. www.tsl.state .tx.us/ld/pubs/crew/index.html

"Leaders and Trendsetters Agree More Students Should Learn Computer Science." *Promote Computer Science*, 2015, https://code.org/promote

Leask, Amy. "Canadian STEM Museums, Coast to Coast." In STEM, STEAM, and STREAM Blog. May 1, 2013. http://enableeducation.com/canadian-stem-museums-coast-to-coast/

Leask, Amy. "20 American STEM Museums You Shouldn't Miss." In STEM, STEAM, and STREAM Blog. May 6, 2013. http://enableeducation.com/20-american-stem-museums-you-shouldnt-miss/

Levy, Joel. *Chemistry in 100 Numbers: A Numerical Guide to Facts, Formulas, and Theories.* London: Apple Press, 2015.

Linz, Ed, Mary Jane Heater, and Lori A. Howard. *Team Teaching Science: Success for All Learners.* Arlington, VA: NSTA Press, 2011.

Loertscher, David. V. "The Virtual Makerspace: A New Possibility." *Teacher Librarian* 43, no. 1 (October 2015).

Lombardi, Abby. "Software Development Ranks as the Most In-Demand Skill for Tech Jobs." *Wanted Analytics,* September 5, 2013. https://www.wantedanalytics.com/analysis/posts/software-development-ranks-as-the-most-in-demand-skill-for-tech-jobs

Lynch, Sharon J., Erin Peters-Burton, and Michael Ford. "Building STEM Opportunities for All." *Educational Leadership* 72, no. 4 (2014): 54–60.

MacNeal, Noel. *Box! Castles, Kitchens, and Other Cardboard Creations for Kids.* Guilford, CT: Lyons Press, 2013.

MacNeal, Noel. *10-Minute Puppets.* New York: Workman Publishing, 2010.

Maeda, John. "STEM to STEAM: Art in K-12 Is Key to Building a Strong Economy." *Edutopia.* October 22, 2012. http://www.edutopia.org/blog/stem-to-steam-strengthens-economy-john-maeda

"Makerspace Materials" appended from "100+ Makerspace Products & Materials." Makerspaces.com. pp. 1–11, accessed December 22, 2016, https://www.makerspaces.com/makerspace-materials-supply-list/

Mardis, Marcia A. *The Collections at the Core: Revitalize Your Library with Innovative Resources for the Common Core and STEM.* Santa Barbara, CA: Libraries Unlimited, 2014.

Markless, Sharon, and David Streatfield. *Evaluating the Impact of Your Library.* 2nd ed. London: Facet Publishing, 2013.

Martinez, Sylvia, and Gary Stager. "The Maker Movement: A Learning Revolution." July 21, 2014. International Society for Technology in Education (ISTE).

Massachusetts Department of Elementary and Secondary Education. "Science and Technology/Engineering Education for All Students: A Vision." 2016. http://www.doe.mass.edu/frameworks/scitech/2016-04/Vision.pdf

Master's in Data Science. "STEM Fun for Kids K-12." http://www.mastersindatascience.org/blog/the-ultimate-stem-guide-for-kids-239-cool-sites-about-science-technology-engineering-and-math/

Matthews, Joseph R. *The Evaluation and Measurement of Library Services.* Westport, CT: Libraries Unlimited, 2007.

Matthews, Joseph R. *Measuring for Results: The Dimensions of Public Library Effectiveness.* Westport, CT: Libraries Unlimited, 2004.

Maxwell, Angela. "The Maker Mindset: Curate, Create, Collaborate." *Ohio Media Spectrum* 67, no. 1 (Fall 2015).

Mayes, Robert, and Thomas R. Koballa, Jr.. "Exploring the Science Framework." *Science Teacher* 79, no. 9 (2012): 27–34. *Academic Search Premier,* EBSCOhost (cited October 7, 2016). https://www.iste.org/explore/ArticleDetail?articleid=106

McCulloch, Catherine. Elevating and Enhancing the "E" in STEM Education. *Learning and Teaching Blog*. June 20, 2016, http://ltd.edc.org/elevating-and-enhancing-engineering-ed

McLaughlin, Molly K. "The Secret to Crowdfunding Success for Inventors and Backers." *PC Magazine Digital Edition*, October 2016. http://web.a.ebscohost .com/ehost/pdfviewer/pdfviewer?sid=31af927c-4821-4607-93eb-427a7142 8be0%40sessionmgr4010&vid=1&hid=4207

Mercer, Bobby. *The Robot Book: Build & Control 20 Electric Gizmos, Moving Machines, and Hacked Toys* (Science in Motion). Chicago Review Press, 2014.

Miaoulis, Ioannes. "K-12 Engineering: The Missing Core Discipline." In Senay Purzer, Johannes Strobel, and Monica E. Cardella (eds.), *Engineering in Pre-College Settings; Synthesizing Research, Policy and Practice*, 35–60. Lafayette, IN: Purdue University Press, 2014.

Miller, R. T. and Girmscheid, L. "It Takes Two." *School Library Journal* 58, no. 5 (May 2012): 25–29. http://search.ebscohost.com/login.aspx?direct=true&db=f5 h&AN=74998359&site=eds-live

Miller, R. T. "We Need Tag-Team Leadership." *School Library Journal* 58, no. 5 (May 2012): 11. http://search.ebscohost.com/login.aspx?direct=true&db=tfh& AN=74998347&site=eds-live

Miller, William, and Rita M. Pellen, eds. *Libraries beyond Their Institutions: Partnerships That Work*. Binghamton, NY: Haworth Information Press, 2005–2006.

Moore, Tamara J., Micah S. Stohlman, Hui-Hui Wang, Kristina M. Tank, Aran W. Glancy, and Gillian H. Roehrig. "Implementation and Integration of Engineering in K-12 STEM Education." In Senay Purzer, Johannes Strobel, and Monica E. Cardella (eds.), *Engineering in Pre-College Settings; Synthesizing Research, Policy and Practice*, 35–60. Lafayette, IN: Purdue University Press, 2014.

Moorefield-Lang, Heather. "Libraries and the Rift: Oculus Rift and 4E Devices in Libraries." *Knowledge Quest* 43, no. 5 (2015): 76–77.

Murvosh, M. "Partners in Success." *School Library Journal* 59(1), (January 2013): 22–28. http://search.ebscohost.com/login.aspx?direct=true&db=tfh&AN= 84641701&site=eds-live

Morgan, Emily, and Karen Ansberry. *Even More Picture-Perfect Science Lesson, K-5: Using Children's Books to Guide Inquiry*. Arlington, VA: NSTA Press, 2013.

Nager, Adams, David M. Hart, Stephen Ezell, and Robert D. Atkinson. "The Demographics of Innovation in the United States." Information Technology and Innovation Foundation. February 24, 2016. https://itif.org/publications/ 2016/02/24/demographics-innovation-united-states

National Center for Literacy Education (NCLE). "Novel Engineering Project Blends Literature, STEM." *NCLE Smart Brief*, December 17, 2015.

National Coalition for Arts Standards. *A Conceptual Framework for Arts Learning*. State Education Agency Directors of Arts Education (SEADAE), 2013. http://www.nationalartsstandards.org/sites/default/files/Conceptual Framework 07–21–16.pdf

National Coalition for Arts Standards. *National Core Arts Standards; Dance, Media Arts, Music, Theater and Visual Arts*. 2014. http://www.nationalartsstandards.org

National Council of Teachers of Mathematics (NCTM). *Principles and Standards for School Mathematics, Grades 9–12 Edition*. Reston, VA: National Council of Teachers of Mathematics, 2000.

National Governors Association. "Building a Science, Technology, Engineering and Math Education Agenda: An Update of State Actions." December 2011, accessed January 21, 2016, http://www.nga.org/files/live/sites/NGA/files/pdf/1112STEMGUIDE.PDF

National Governors Association Center for Best Practices, Council of Chief State School Officers. *Common Core State Standards*. Washington, DC: National Governors Association Center for Best Practices, Council of Chief State School Officers, 2010. http://www.corestandards.org

National Research Council (NRC). *A Framework for K–12 Science Education: Practices, Crosscutting Concepts, and Core Ideas*. Washington, DC: National Academies Press, 2012.

National Science Teacher Association (NSTA). *Access the Next Generation Science Standards by Topic*. ngss.nsta.org/AccessStandardsByTopic.aspx

National Science Teachers Association (NSTA). "Best STEM Books." 2016–2017. http://static.nsta.org/pdfs/2017BestSTEMBooks.pdf

National Science Teacher Association (NSTA). *Next Generation Science Standards: For States, By States* (2013). Washington, DC: The National Academies Press. www.nextgenscience.org/get-to-know

National Science Teachers Association (NSTA). "NGSS@NSTA STEM Starts Here." http://ngss.nsta.org

National Science Teachers Association (NSTA). "Science Resources for Parents." 2017. http://www.nsta.org/parents/

NGSS Lead States. *Next Generation Science Standards: For States, By States*. Washington, DC: The National Academies Press, 2013.

NMC. COSN. "NMC/CoSN Horizon Report 2016 K-12 Edition." http://cdn.nmc.org/media/2016-nmc-cosn-horizon-report-k12-EN.pdf

November, Alan. "Clearing the Confusion Between Technology Rich and Innovative Poor: Six Questions." November Learning. January 12, 2015. http://novemberlearning.com/assets/ClearingtheConfusionbetweenTechnologyRichandInnovativePoorSixQuestions.pdf

November, Alan. "The Seven Questions Every New Teacher Should be Able to Answer; #1." Blog. June 13, 2016 (reprinted December 30, 2016). p. 2. http://www.eschoolnews.com/2016/12/30/1-questions-new-teacher-answer/

Noyce Foundation. Examining the impact of afterschool STEM programs. July 2014. http://www.afterschoolalliance.org/ExaminingtheImpactofAfterschoolSTEMPrograms.pdf

Obama, Barack. "Remarks by the President at White House Science Fair," March 23, 2015, In U.S. Department of Education. "Science, Technology, Engineering and Math: Education of Global Leadership," accessed January 21, 2016, http://www.ed.gov/stem

O'Clair, Katherine, and Jeanne R. Davidson, eds. *The Busy Librarian's Guide to Information Literacy in Science and Engineering*. Chicago: ACRL, 2012.

Office of the President of the United States, President's Council of Advisors on Science and Technology. *Engage to Excel: Producing One Million Additional College Graduates with Degrees in Science, Technology, Engineering, and Mathematics*, 2012.

"100+ Makerspace Products & Materials." Makerspaces.com:1–11, accessed December 22, 2016, https://www.makerspaces.com/makerspace-materials-supply-list/

O'Reilly, Katie. "Libraries on Lockdown: Escape Rooms, a Breakout Trend In Youth Programming." *American Libraries*, 47, no. 9/10 (September/October 2016): 14–17.

Oremland, Sara. "Collaboration and Technology for Authentic Research Projects: From Essential Question to Presentation." *Knowledge Quest* 41, no. 4 (2013): 60–68.

Pandora, Cherie P. and Stacey Hayman. *Better Serving Teens through School Library— Public Library Collaborations*. Santa Barbara, CA: Libraries Unlimited, 2013.

Peterson, Tommy. "3D Printing Adds Another Dimension to the Classroom". 2015. *EdTech Magazine*. http://www.edtechmagazine.com/k12/article/2015/01/new-dimension

Perez, Lisa. "Master Librarian: Mentoring Teachers to Win the Technology Wars." *Knowledge Quest* 41, no. 4 (2013): 22–26.

PEW Research Center. "Minorities Account for Nearly All U.S. Population Growth." FACTANK: News in the Numbers. March 30, 2011, p. 1. http://www.pewresearch.org/fact-tank/2011/03/30/minorities-account-for-nearly-all-u-s-population-growth/

Pierce, Jennifer Burek. "Your Story Matters." *American Libraries* 47, no. 1–2 (2016): 82.

Pink, Daniel H. A *Whole New Mind; Why Right-Brainers with Rule the Future*. New York: Riverhead Books, 2006.

Pittinsky, Todd L., and Nicole Diamante. "Going Beyond Fun in STEM." *Phi Delta Kappan*, 97, no. 2 (2015): 47–51.

Plass, Jan. "Why Games and Learning." Institute of Play.org. http://www.instituteofplay.org/about/context/why-games-learning/

Pledger, Marcia. "Manufacturing contest [M]SPIRE seeks online applicants for grant money." *The Plain Dealer* (Cleveland, OH) August 26, 2016. http://www.cleveland.com/business/index.ssf/2016/08/manufacturing_contest_mspire_s.html

Pomeroy, Steven Ross. "From STEM to STEAM: Science and Art Go Hand-in-Hand." Blog. *Scientific American*, August 22, 2012, accessed January 21, 2016, http://blogs.scientificamerican.com/guest-blog/from-stem-to-steam-science-and-the-arts-go-hand-in-hand/

Poot, Astrid. "Making Makers." (2016). http://lekkersamenklooien.nl/wp-content/uploads/2017/02/makersmaken_50tools_ENG.pdf

Poot, Astrid. "Unlock the Maker." (Button 2017). http://lekkersamenklooien.nl/

Porter, Marcia. Books, Bytes. Blog. "Lego Desk." January 7, 2017. Common Ground 2016 Presentation.

Preddy, Leslie B. *School Library Makerspaces: Grades 6–12*. Santa Barbara, CA: Libraries Unlimited, 2013.

President's Committee on the Arts and the Humanities. *Reinvesting in Arts Education: Winning America's Future through Creative Schools*. Washington, DC: President's Committee and the Arts and the Humanities, 2011. http://www.pcah.gov/sites/default/files/photos/PCAH_Reinvesting_4web.pdf

Professional Values Task Force. Young Adult Services Association. *Core Professional Values for the Teen Services Profession*. Chicago: American Library Association, 2015.

Purcell, Melissa A. *The Networked Library: A Guide for the Educational Use of Social Networking Sites*. Santa Barbara, CA: Libraries Unlimited, 2012.

Raising Little SuperHeroes. "May the Fourth Be With You Light Saber Card." 2017. Blog. http://raisinglittlesuperheroes.com/lightsaber-card/

Ramirez, Ainissa. "The Immutable Impact of Black Scientists and Inventors." *Edutopia: Technology Integration*. Blog. February 12, 2014. https://www.edutopia.org/blog/impact-of-black-scientists-inventors-ainissa-ramirez

Rendina, Diana. "4 Super Easy Budget Friendly Projects for Your Makerspace." December 19, 2016. Blog. http://renovatedlearning.com/2016/12/19/budget-friendly-projects-makerspace/

Rendina, Diana. "How to Start a Makerspace When You're Broke." *Knowledge Quest* Online. February 22, 2016. http://knowledgequest.aasl.org/start-makerspace-youre-broke/

Rendina, Diana. Lowe's Toolbox for Education Grant. Toolboxforeducation.com. Blog. "Library Makeover Preview." May 23, 2014.

Rendina, Diana. Presentation at AASL: *Makerspaces and Libraries: How to Bring Some STEAM into Your Program*. Columbus, OH, 2015.

Rendina, Diana. "STEM Maker List Grant." Blog. July 30, 2014. http://renovatedlearning.com/

Rendina, Diana. "Why a Makerspace Is Not a Magic Cure-All for Your Problems." AASL. *Knowledge Quest* Blog, December 29, 2016.

Riley, Susan. "Is It Arts Integration or STEAM?" *Education Closet*. November 30, 2016. https://educationcloset.com/2016/11/30/arts-integration-steam/

Root-Bernstein, Robert, and Michele Root-Bernstein. "The Art & Craft of SCIENCE." *Educational Leadership* 70, no. 5 (February 2013): 16–21.

Root-Bernstein, Robert S., and Michele M. Root-Bernstein. *Sparks of Genius*. New York: Mariner Books, 2001.

Rossman, Edmund A. *40+ New Revenue Sources for Libraries & NonProfits*. Chicago: ALA, 2016.

Rothwell, Jonathan. "The Need for More STEM Workers." *The Avenue*. The Brookings Institution, June 1, 2012. https://www.brookings.edu/blog/the-avenue/2012/06/01/the-need-for-more-stem-workers/

Royce, Christina. *Teaching Science through Trade Books*. Arlington, VA: National Science Teachers Association, NSTA Press, 2012.

Ruhlmann, Ellyn. "Connecting Latinos with Libraries". *American Libraries* 45, no.5 (May 2014): 36–40.

Samson, Ghislain. "From Writing to Doing: The Challenges of Implementing Integration (and Interdisciplinarity) in the Teaching of Mathematics, Sciences, and Technology." *Canadian Journal of Science, Mathematics & Technology*. 14, no. 4 (October-December 2014): 346–358.

Schley, Courtney. "Learning & STEM Toys We Love: Everything from Games to Maker Kits, Robots, and Crafts." *The Wirecutter in Gadgetry*, December 2, 2016.

Schwartz, Katrina. "Can Micro-credentials Create More Meaningful Professional Development for Teachers?" *Mindshift*. KQED. February 15, 2017. https://ww2.kqed.org/mindshift/2017/02/15/can-micro-credentials-create-meaningful-professional-development-for-teachers/

Scieszka, Jon. *The Math Curse*. New York: Viking, 1995.

Scordato, Julie, and Ellen Forsyth, eds. *Teen Games Rule! A Librarian's Guide to Platforms and Programs*. Santa Barbara, CA: Libraries Unlimited, 2013.

Scripa, Allison, and Heather Moorefield-Lang. "Putting the Citizen in Science." *Knowledge Quest* 41, no. 4 (2013): 54–59.

Search Institute. "40 Developmental Assets for Adolescents (ages 12–18)." 2007. http://www.search-institute.org/system/files/40AssetsList.pdf.

"7 Things You Should Know About Makerspaces." *EDUCAUSE Learning Initiative*. April 9, 2013, accessed January 21, 2016, http://www.educause.edu/library/resources/7-things-you-should-know-about-makerspaces

Seymour, Gina. "STEAM + C: Adding Compassion to the Makerspace 'Because Nice Matters'." Blog. January 25, 2017. https://ginaseymour.com/

Shetterly, Margot Lee. *Hidden Figures*. New York: HarperCollins, 2016.

Silver, Kate. "Get Cracking on Code: Community Courses Lead to Jobs." *American Libraries* 46, no. 3–4 (2015): 56–59.

Slaughter, Louise. "Proof That the STEM Fields Are Totally Inhospitable to Women." *MS Magazine* Blog. April 14, 2016. http://msmagazine.com/blog/2016/04/14/proof-that-the-stem-fields-are-totally-inhospitable-to-women/

Smallwood, Carol, ed. *Bringing the Arts into the Library*. Chicago: American Library Association, 2014.

Smallwood, Carol, and Kim Becnel, eds. *Library Services for Multicultural Patrons: Strategies to Encourage Library Use*. Lanham, MD: Scarecrow Press, 2013.

Snyder, Maureen M. and Roche, Janet. "Road Map for Improvement Evaluating Library Media Programs." *Knowledge Quest* 37, no. 2 (November/December 2008): 22–27.

Song, Ting, and Kurt Becker. *Technology and Engineering Teacher* 73, no. 2 (October 2013): 30–34.

Southorn, Graham, and Giles Spar. *Physics Squared: 100 Concepts You Should Know*. London: Apple Press, 2016.

Squires, Tasha. "Engaging Students through Gamification." *American Libraries* 47, no. 3/4 (March/April 2016): 20.

Squires, Tasha. *Library Partnerships: Making Connections between School and Public Libraries*. Medford, NJ: Information Today, 2009.

Stephan, Michelle. "Off to the Duck Races: Planning for Inquiry in STEM." *Educational Leadership* 74, no. 2 (October 2016). http://www.ascd.org/publications/educational-leadership/oct16/vol74/num02/Off-to-the-Duck-Races@-Planning-for-Inquiry-in-STEM.aspx

Stewart, Becky. "Challenging Perceptions in the STEM Classroom." NSTA Blog. February 21, 2015. http://nstacommunities.org/blog/2015/02/21/challenging-perceptions-in-the-stem-classroom/

Stripling, Barbara K. "Equity, Diversity, and Inclusion." *American Libraries* 45, no. 5 (May 2014): 5.

Stripling, Terri, and Beverly Simmons. "Get Students Revved Up! Robotics Bring Excitement to STEM." *Tech Directions* 75, no. 7 (March 2016): 13–17.

Stuart, Colin. *Math in 100 Numbers: A Numerical Guide to Facts, Formulas, and Theories*. London: Apple Press, 2015. (Currently unavailable)

Stuart, Colin. *Physics in 100 Numbers: A Numerical Guide to Facts, Formulas, and Theories*. London: Apple Press, 2015.

Sullivan, Margaret L. *High Impact School Library Spaces: Envisioning New School Library Concepts.* Santa Barbara, CA: Libraries Unlimited, 2014.

"SWOT Analysis: Strengths, Weaknesses, Opportunities and Threats." 2014. https://eclkc.ohs.acf.hhs.gov/hslc/tta-system/operations/mang-sys/planning/2SWOTAnalysisS.htm

"The Tech Challenge 2017: Rock the Ravine." The Tech Museum of Innovation. 2017. https://www.thetech.org/thetechchallenge

Todaro, Julie Beth. "Community Collaborations at Work and in Practice Today: An A to Z Overview." In William Miller and Rita M. Pellen, eds. *Libraries beyond Their Institutions: Partnerships That Work.* Binghamton, NY: Haworth Information Press, 2005–2006. 137–156.

Tsupros, N., Kohler, R., & Hallinen, J. *STEM Education: A Project to Identify the Missing Components.* Carnegie Mellon University, PA: Intermediate Unit 1: Center for STEM Education and Leonard Gelfand Center for Service Learning and Outreach, 2009.

2016 Boy Scout Requirements. Irving, TX: Boy Scouts of America, 2016.

U.S. Census Bureau. "Census Bureau Reports Majority of STEM College Graduates Do Not Work in STEM Occupations." July 10, 2014.

U.S. Department of Education. "Science, Technology, Engineering and Math: Education of Global Leadership, Five-Year Strategic Plan for STEM Education," accessed January 21, 2016, http://www.ed.gov/stem

Varnum, Kenneth J. *The Top Technologies Every Librarian Needs to Know.* Chicago: American Library Association, 2014.

Vasquez, Jo Anne. "STEM Beyond the Acronym." *Educational Leadership* 72, no. 4 (December 2014/January 2015): 10–15.

Vasquez, Jo Anne, Cary Schneider, and Michael Comer. *STEM Lesson Essentials: Integrating Science, Technology, Engineering and Math, Grades 3–8.* Arlington, VA: NSTA Press, 2013.

Vilorio, Dennis. "Stem 101: Intro to Tomorrow's Jobs." *Occupational Outlook Handbook.* (2014): 2–12. http://www.bls.gov/careeroutlook/2014/spring/art01.pdf

Wall, Cindy R. and Lynn M. Pawloski. *The Maker Cookbook: Recipes for Children's and 'Tween Library Programs.* Santa Barbara, CA: Libraries Unlimited, 2014.

Wallace, Danny P. *Library Evaluation: A Casebook and Can-Do Guide.* Santa Barbara, CA: Libraries Unlimited, 2001.

"What Is A Badge?" (video) 2:44. Mozilla Foundation. https://youtu.be/HgLLq7ybDtc

Wheeler-Toppen, Jodi, and Carol Tennant. *Edible Science: Experiments You Can Eat.* Washington, DC: Ivy Press Limited, National Geographic Society, 2015.

Whittingham, Jeff, and Wendy A. Rickman. "Booktalking: Avoiding Summer Drift." *Knowledge Quest* 43, no. 5 (2015): 18–21.

Williams, Joseph P. "Expanding the STEM Pipeline." *U.S. News Digital Weekly* 6, no.23 (2014): 13.

"Women in STEM: A Gender Gap to Innovation." Washington, DC: U.S. Department of Commerce, Economics and Statistics Administration, August 3, 2011. http://www.esa.doc.gov/sites/default/files/womeninstemagaptoinnovation8311.pdf

Wright, Joshua. "America's Skilled Trades Dilemma: Shortages Loom as Most-In-Demand Group of Workers Ages." *Forbes*. March 7, 2013. https://www.forbes.com/sites/emsi/2013/03/07/americas-skilled-trades-dilemma-shortages-loom-as-most-in-demand-group-of-workers-ages/#fbb9b64 6397c

Xue, Yi, and Richard C. Larson. "STEM crisis or STEM surplus? Yes and yes." U.S. Bureau of Labor Statistics, May 2015. https://www.bls.gov/opub/mlr/2015/article/stem-crisis-or-stem-surplus-yes-and-yes.htm

Yager, Robert E., and Herbert Brunkhorst. *Exemplary STEM Programs: Designs for Success*. Arlington, VA: NSTA Press, 2014.

Yerak, Becky. "STEM-inspired Dolls in Demand." *The Plain Dealer* (Cleveland, OH). December 20, 2015.

Young Adult Library Services Association (YALSA). "STEM Resources." http://wikis.ala.org/yalsa/index.php/STEM_Resources

"Youngstown STEM Workshop Targets Girls." *Vindicator* (Youngstown, OH), July 05, 2013. *Newspaper Source*, EBSCOhost.

Zalaznick, Matt. 2015. "Putting the 'A' in STEAM." *District Administration* 51, no. 12 (2015): 62–66.

INDEX

ABOUT THE AUTHORS

CHERIE P. PANDORA served as a teacher/school librarian for 35 years, taught research at a career college, wrote grants, and was awarded the Ohio Educational Library Media Association (OELMA) Award of Merit in 2013 for contributions to School Librarianship. She received Ohio Master Teacher status and her library was recognized as an OELMA Library of Distinction. Her published work includes Libraries Unlimited's *Better Serving Teens through School Library–Public Library Collaborations*. She has presented nationally at conferences of the American Association of School Librarians (AASL) and the Public Library Association (PLA). In Ohio, she has made presentations at eTech, OELMA, the Academic Library Association of Ohio (ALAO), and Kent State University workshops and has written frequently for *Ohio Media Spectrum*. Pandora earned an Educational Specialist degree (staff and personnel) from Cleveland State University and a Master of Library Science degree from Kent State University.

KATHY FREDRICK has been a school librarian, district library director, and technology director over her 40-year career in Wisconsin, Ohio, Australia, and Germany. She has presented at conferences such as Ohio Educational Media Association (OELMA) and Ohio eTech, and worked on programs for teachers and administrators on technology integration, digital citizenship, and using online resources. Fredrick wrote a column for *School Library Monthly* (now *School Library Connection*) for eight years focused on emerging technologies and the school library. She holds a Masters of Library and Information Science from the University of Wisconsin-Madison and an Education Specialist license in Curriculum and Instruction from Cleveland State University. Fredrick has been recognized for her work by the WVIZ/PBS Educational Advisory Board and the Cleveland Area Metropolitan Library System (CAMLS), a regional multitype library consortium.